Ptolemy
The Book of
Astronomy in
Antiquity

the building
blocks of life

foundations

science

This is a **FLAME TREE** book

FLAME TREE PUBLISHING
6 Melbray Mews, Fulham,
London SW6 3NS, United Kingdom
www.flametreepublishing.com

First published and copyright
© 2024 Flame Tree Publishing Ltd

This edition is an abridged version based on *Ptolemy's Almagest* by
Claudius Ptolemy, translated and annotated by G.J. Toomer, 1998
(Princeton University Press, New Jersey).

24 26 28 27 25
1 3 5 7 9 10 8 6 4 2

ISBN: 978-1-80417-791-4

Cover and pattern art was created by Flame Tree Studio.

A copy of the CIP data for this book is available
from the British Library.

Designed and created in the UK | Printed and bound in China

Ptolemy
The Book of
Astronomy in
Antiquity

A New Introduction by Christián C. Carman

the building
blocks of life

foundations

science

Series Foreword by Professor Marika Taylor

FLAME TREE
CONCISE
EDITIONS

Contents

Series Foreword

S cience has always been a source of fascination, pushing the frontiers of knowledge. This series of books spans some of the most important scientific developments of all time. From the origins of life to the nature of the universe, remarkable achievements through the ages are showcased, as well as the limitless potential of human curiosity.

In this series we can follow the evolution of science and philosophy from ancient times through to the modern era. The works of Plato, Ptolemy and Aristotle cover a wide range of topics related to the natural world, including physics, metaphysics and biology. Aristotle's work is particularly notable for its depth and breadth as well as for its immense influence on the development of scientific and philosophical thought in the Western world.

Copernicus and Newton were also great polymaths. Copernicus is best known for his work on the heliocentric nature of the solar system as presented in *On the Revolutions of the Heavenly Spheres* but he also made substantial contributions to mathematics and medicine. Newton's *Principia* is recognized as one of the most important scientific works ever written. His theories of motion and gravity laid the foundations for modern physics and revolutionized our understanding of the 'Nature' of the world. Within mathematics, Newton developed calculus, as well as algebra, trigonometry and geometry.

The *Principia* was originally read primarily by intellectuals. By contrast, Darwin's work *On the Origin of Species* immediately caused a significant stir in both scientific and public circles, particularly through its challenge to traditional beliefs. Darwin's theory of evolution was debated and refined, gradually becoming a foundational concept of modern biology.

At the beginning of the twentieth century, physics experienced two major paradigm shifts: relativity and quantum theory. Einstein's *Special and General Relativity* fundamentally changed our understanding of space, time and gravity. Bohr and Planck explain the early development of quantum theory, starting from Planck's revolutionary concept of quanta. In the same period, pioneering new fields of research were established all across science, including Freud's renowned psychoanalysis.

Readers of this series can step back in time to explore how the foundations of science and culture are described by its pioneers. The transformative nature of these works and their profound impacts on society can only inspire a sense of awe and wonder.

Professor Marika Taylor
University of Southampton

A New Introduction

Astronomy as Therapy
in Ancient Times

I n Plato's *Republic*, Socrates discusses with Glauco, his disciple, every detail of an ideal republic. When it comes to education, Socrates and Glauco ponder about what the disciplines are that every one of its citizens should study. Socrates proposes astronomy as a candidate. Glauco agrees that everyone should study astronomy. He adds his reasons: it is necessary for agriculture and navigation. Of course, in ancient times, astronomy was essential for measuring time and knowing when to reap and when to sow; it was also used by those at sea to orientate themselves by studying the stars. Today, equipped with a watch and GPS, we do not need it in the same way.

Faced with Glauco's response, however, Socrates laughs and gently teases his disciple: 'it seems as if you are afraid that students will reproach you for proposing useless studies'. Astronomy, Socrates continues, does not only need to be studied because it is useful. It has a much deeper, therapeutic function. Socrates explains that excessive pleasures and obsession with everyday occupations end up distorting the soul. 'Thanks to the study of astronomy, the soul of each person is purified and revived when it is dying and blinded by other occupations.'

Astronomy helps you to switch attention from the chaotic, ever-changing concerns of everyday life to the everlasting stars, dancing in an eternal harmony and unalterable peace. Socrates tells Glauco that astronomy and music are twinned: just as ears were created for music, so were eyes for astronomy. Studying the stars serves to connect the soul with the most enduring, with the spark of eternity that is in us.

This cathartic perception to astronomy did not only reflect the romantic vision of philosophers. Professional astronomers of the highest level also shared this view. A beautiful poem by Claudius Ptolemy (c.100–c.165 CE), the greatest astronomer of antiquity and one of the most influential scientists in history, manifests exactly the same in expressing what he feels about the discipline:

11

I know that death marks my days / but when I contemplate the unceasingly rotating stars / my feet no longer stand on Earth / and, next to Zeus himself, / the part of immortality that belongs to me I claim.

For Ptolemy, astronomy was part of theology, the study of the gods. Such a profound subject prompted the ancients to develop a far more precise conception of astronomy than would have been required for mere navigation or the proper management of planting and harvesting.

The Life and Works of Claudius Ptolemy

Virtually nothing is known about the life of Claudius Ptolemy. The best explanation is probably that he was simply a great scientist, and nothing else, so he did not arouse much interest among ancient writers. From his works, we can only infer the approximate time and place in which he worked. The first observation that Ptolemy made that is recorded in the *Almagest* is from 5 April 125, in Alexandria, where he reported a lunar eclipse; the last one he mentions, also in Alexandria, is from 2 February 141, where he recorded the position of Mercury with respect to the Sun. We also know that he wrote many works after the *Almagest*, some really important and extensive, so he must have continued to be active in his written work for at least another 20 years.

We can therefore conjecture that he was born around the year 100 and died around the year 165, perhaps a little later. We do not know precisely where he was born; his name, 'Claudius', is clearly Greek, while 'Ptolemy' may indicate that he originated from one of the Egyptian towns named after the kings of the Ptolemaic dynasty. What is clear is that his link with the Ptolemaic kings, leading to his appearance in royal clothing and insignia in some paintings, most notably Raphael's *School of Athens*, is nothing more than a confusion that lasted until the beginning of modernity. From the fact that Ptolemy refers to many observations made from Alexandria, and that he had access to observational registers only available in a large library, we can say with confidence that he worked in that great Egyptian city. It is possible that he lived in the nearby town of Canopus. Luckily at least his poem (see opposite) has been preserved, enabling us to penetrate a little into the depth of his soul.

Although Ptolemy is known as a great astronomer, a quick review of his works shows that his interests were far more universal. His works could form a kind of ancient 'Encyclopedia of Applied Mathematics', in contrast to already existing works dedicated to pure mathematics, such as Euclid's *Elements*. In addition to astronomy, Ptolemy wrote on geography, climatology, optics and harmonics; he also produced some philosophical and astrological works.

The *Almagest*

The *Almagest* is, without a doubt, Ptolemy's greatest astronomical work; it is also one of the most wonderful scientific works of all time. Its English name derives from *Almagestum*, the Latin translation of the name by which the book became popularised in the Arab word, *Kitab al-majistī*. 'Kitab' means book, but the meaning of al-majistī is not so clear. The Greek name of the work is Μαθηματικὴ Σύνταξις (Mathēmatikē Syntaxis), literally meaning 'Mathematical Collection'. There is evidence that it was later called, simply, Ἡ Μεγάλη Σύνταξις (Hē Megalē Syntaxis) – 'The Great Collection'. The superlative of Μεγάλη (Megalē) is μεγίστη (megistē), i.e., 'the greatest'. Most likely, the Arabs added the article 'al', thus creating the al-megiste, which simply means 'The Greatest'. Anyone who has investigated the work will know that such a name is not disproportionate.

The Structure of the *Almagest*

The structure of the work, which consists of 13 'books' (we would say 'chapters' today), is explained by Ptolemy himself at the beginning of the first book, after a preface of philosophical content. First of all, it features a brief discussion about the Earth: its position at the centre of the universe and its immobility. This is followed by a longer one, instrumental, in which geometric solutions to problems related to spherical astronomy are exposed (the end of Book I and the entire Book II). Next the models of the Sun and the Moon are developed (Books III to VI) and then the theory of the stars – first that of the fixed ones (Books VII and VIII) and finally that of the five planets (Books IX–XIII). The structure of the work follows a strict axiomatic order: it is through the position of the Sun that that of the Moon can be determined, and through the position of both the Sun and Moon that the positions of the stars can be obtained. Finally, the planets are usually located by referring to stars.

· Fortunately Ptolemy was very scrupulous in quoting his predecessors when an idea was not his own. This makes it possible for us clearly to distinguish which are, and which are not, his own contributions. The solar theory, the fixed stars and part of the lunar model are inherited by Ptolemy from previous astronomers, mostly from Hipparchus of Nicaea (*c*.190–*c*.120 BCE). Nevertheless, he manages to systematize them harmoniously, integrating them into a single system. In the case of the Moon, in addition, Ptolemy proposes very important modifications to the pre-existing models. However, his originality is most manifest, without a doubt, in a planetary theory of such predictive accuracy that it would only be surpassed 1,500 years later, by the introduction of the planetary laws of Johannes Kepler (1571–1630 CE).

A Note on this Abridgement

This volume takes its text from the 1998 Princeton edition of G.J. Toomer's translation of the *Almagest*, so the reader must please note that any reference to sections of the full version of the text applies to that edition. For our abridgement, the parts selected are the ones that will be of greatest interest to a non-specialist reader, including those parts in which Ptolemy's originality stands out, while those that are more technical have been left out. Not included are: the analysis of how to obtain values for trigonometric functions (second part of Book I), the developments of spherical astronomy (Book II), the very long catalog with more than 1000 stars (Book VII), the theory of planetary latitudes (Book XIII), and numerous tables that accompany the planetary models. Instead, we have included the philosophical introduction, the discussion about the immutability and centrality of the Earth (the first part of Book I), the fundamentals of the theory of the Sun (Book III), of the lunar theory (parts of Books IV and V), of the stars (beginning of Book VII) and of planetary theories (Books IX and X). This selection allows us to appreciate Ptolemy's great creative and systematizing capacity, without getting lost in calculations and technicalities.

Please note too that, in order to avoid confusion in comparison with the full edition, the footnotes in our edition run on from the start, rather than beginning afresh with each Book.

The Sphericity of the Earth

From at least since the times of the pre-Socratic philosophers, the Earth was considered to be a sphere. Any educated person during ancient and medieval times believed the Earth to be spheric. The idea that the sphericity of the Earth was still the subject of debate in early modern times, and that Chistopher Colombus (1451–1506 CE) tried to prove its sphericity by circumnavigating the globe, is a myth. In his book *Inventing the Flat Earth*, Jeffrey B. Russell has analysed the origin and causes of this fable in detail. He shows that it is based on a campaign to discredit the Middle Ages. In any case Aristotle (384–322 BCE), in his work *On the Heavens*, develops several arguments in favour of the sphericity of the Earth. The most eloquent is that of the lunar eclipses.

We know that in a lunar eclipse, the shadow of the Earth is projected onto the lunar surface. The Earth's shadow is always circular, and the sphere is the only shape that always projects a circular shadow. Ptolemy added other arguments to those of Aristotle. Some are more technical, but he also mentions the argument usually attributed to Colombus: if we sail towards the mountains

they are observed to increase gradually in size and as if rising up from the sea itself in which they had previously been submerged: this is due to the curvature of the surface of the water.

The Sphere of the Fixed Stars

It was from a spherical Earth, therefore, that ancients looked at the sky. Astronomers in those times did not use telescopes or other sophisticated optical instruments, so what they saw when looking at the sky is what any of us can see if we look up on any non-cloudy night, given a sufficiently dark location. If we carefully observe the sky, we will see that all the stars maintain their relative distances, appearing to form an immense vault over our heads. This is why they were called 'fixed stars'. Now that stars were fixed, i.e., that they kept their relative angular separation over time, was not something dogmatic for ancient astronomers. On the contrary: this statement was subjected to empirical scrutiny several times throughout history.

Ptolemy recognizes that the relative position of the stars could change so slowly that individual lifetimes are not enough to perceive the variation. Consequently, he compared the position of several stars with the positions that Hipparchus reported around two centuries before him and asserted that he could not perceive any change. Nevertheless, he adds new data so that future astronomers could continue to evaluate this statement. His effort was certainly not in vain because, 16 centuries later, Edmund Halley (1656–1742 CE) used Ptolemaic observations to detect, for the first time in history, the proper motion of some stars. Their motion is so slow that we needed to compare accurate observations more than 15 centuries apart to detect it.

Their relative motion, therefore, is imperceptible. However, if we observe the sky several times during one night, we will notice that all the starts together rotate in an east–west direction, turning once a day. Even the Sun, the Moon and the planets accompany the stars in this daily revolution (although they also move slowly relative to the stars). This daily motion can be explained in two ways: either everything revolves around the Earth, making one revolution per day from east to west, or the Earth rotates once per day in the opposite direction. In fact Aristarchus of Samos (*c*.310–*c*.230 BCE), among other ancient astronomers, proposed not only the daily rotation of the Earth, but also its annual revolution around the Sun.

The Immobility of the Earth (No Rotation)

Ptolemy is aware that, from a strictly astronomical point of view, what we observe in the sky can be explained by moving the whole sphere of the stars or by moving the Earth (or even by moving both!). However, he adds, the situation is quite different when we consider what happens on the Earth and in the air. If the Earth rotated once a day, it would be impossible for us to see the birds or the clouds moving in the direction of the Earth's rotation. The rotation of the Earth would always overtake that of the birds or clouds, and we would necessarily see them move backwards. We would also always experience a strong wind towards the east.

These arguments were repeated over the centuries until Galileo's refutations. Even if the arguments seem a bit naïve from our modern point of view, what is remarkable is that Ptolemy considered the possibility of the rotation of the Earth and discarded it on *rational* grounds. As in the case of the immobility of the stars, the reasons were not dogmatic, but rational.

19

The Centrality of the Earth

Furthermore, the Earth is at the centre of the sphere of the fixed stars. Ptolemy evaluates different possible positions of the Earth with respect to the sphere of the fixed stars; for each one, he shows consequences that contradict observations. The easiest to see is the following: if we look from any point on Earth, at any time of the year, we will see that the horizon divides the celestial sphere in half. We always see exactly half the sky. That we always see exactly six signs of the zodiac shows it clearly. This is only possible, Ptolemy asserts, if we find ourselves in the middle of that sphere. If not, the horizon would split the sphere into unequal parts and we would not be able to observe half the sky, as the figure below illustrates.

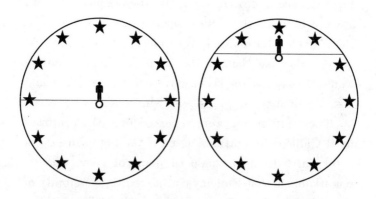

If the Earth is at the centre of the sphere, the observer will see half of the sky; if not, the observer will see less than half.

The Immobility of the Earth (No Revolution)

Drawing on this same argument, Ptolemy refutes the hypothesis of the annual revolution of the Earth around the Sun that Aristarchus has proposed. If the Earth indeed revolves around the Sun, it must at some point leave the centre of the celestial sphere. At that point, we could no longer see exactly half the sky. In reality, however, we always see half of the sky, so the Earth never leaves the centre. Accordingly, it must always rest at the centre.

The Ptolemaic Paradigm

Like any other scientist, Ptolemy works in a scientific framework that establishes some undisputable principles, defines the problems to be solved and provides instruments to deal with them. This framework was popularized by Thomas Kuhn (1922–1996), the twentieth-century philosopher of science, with the concept of 'paradigm'. The paradigm in which Ptolemy worked was characterized by Plato (c.427–347 BCE). In his book, *On the Heavens*, Simplicius (490–560 CE) says that Plato posed the following question to astronomers: 'By hypothesizing which circular and uniform motions will the phenomena of the motions of the planets be saved?' This question has defined the bases of every astronomical research programme for centuries.

Every astronomer, at least until Kepler, understood that the task of astronomy is to look for

21

a combination of circular and uniform motions able to reproduce the non-regular apparent motions of the planets. The undisputable principle is, therefore, that the planetary motion is circular and uniform, the problem to be solved is the apparently irregular motion of the celestial bodies and the most popular instrument available was the combination of circular and uniform motions in a device called an 'epicycle and deferent system'.

Below I will briefly describe the problem to be solved, followed by the solutions proposed.

The Greatest Challenge
to Ancient Astronomy

We already know that the fixed stars revolve around the Earth, keeping their relative positions. One can imagine the stars as fixed inside a sphere, rotating around its axis one revolution per day. The stars in their motion drag the Sun, Moon and the planets that, consequently, also revolve around the Earth. However, they move slower than the stars. Accordingly, they revolve in the contrary direction to the background of the fixed stars (i.e., from west to east). In the case of the Sun, it makes one turn per year, tracing a maximum circle called 'ecliptic' on the sphere of the fixed stars. In addition, the Moon and planets revolve in the contrary direction of the stars – each at its own speed and in a particular plane, but all reasonably close to the plane of the ecliptic. On average, it takes about 27.5 days for the Moon to complete its revolution, for Mercury and Venus exactly one year, for Mars just under two years, for Jupiter just under 12 years and for Saturn almost 29.5 years. I said 'on average' because the Sun, Moon and planets move in a non-uniform motion in their turn around the zodiac. This phenomenon constitutes a challenge to the Platonic request of uniform motion.

However, the most challenging feature for astronomers is the retrograde motion of the planets. From time to time they start to slow down in their motion relative to the stars until they come to a complete stop. They then start to move backwards (i.e., in the direction of daily motion of the fixed stars) for a given time, then stop again and resume their usual direction. If one draws the position of the planets on the background of the fixed stars during the retrograde motion, it produces loops such as those shown in the diagram overleaf. Each planet retrogrades at its own rate – for example, Venus once every 1.6 years and Mars every 2.15 years.

These periods are again mean values because, even for the same planet, the time between one retrogradation and the next varies. Moreover, the size of the loops varies too. For example, in the case of Mars, the smallest loop is around 10° wide, while the biggest doubles it, reaching 20° wide. The attempt to reduce such an apparently chaotic and complex motion to a combination of uniform and circular motions is the story of Western astronomy from Plato to Kepler. In this story, Ptolemy's *Almagest* is the main character.

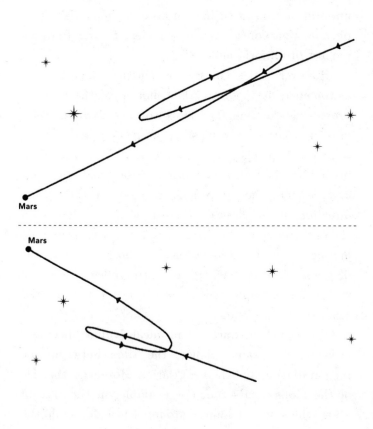

Mars's retrograde loops.

The Epicycle and Deferent System

The geometrical device used by Ptolemy in the *Almagest* to face this challenge was already available to astronomers of his time: the epicycle and deferent model. This device combines two uniform and circular motions in a smart way to produce the retrograde loops. Look at diagram A of the next figure. Here O is the centre of the Earth. Point C revolves in the direction of the zodiacal signs uniformly around O (counterclockwise in the figure) in a circle called 'deferent'. In turn, P revolves around C in the same direction but four times faster than C around O in a circle called the 'epicycle'. P is the planet.

The combination of both uniform and circular motions produces the trajectory in grey that shows the desired loops. When the planet P crosses the dotted lines and is in the inner part of the loop, it seems to retrograde from the observer, inverting its direction. Therefore, in the thin sectors between the dotted lines, the planet is moving backwards.

If one correctly determines the sizes of the circles and the speed of the points revolving on them, the epicycle and deferent model produces the correct number of retrogradations. But it is not so successful in predicting the position of the planet in the sky. In the case of Mars, for example, the error is sometimes enormous, reaching 30°. The main reason is that while the retrograde loops are not distributed uniformly around the zodiac and their loops are not equal in length, as already mentioned, the epicycle and deferent model cannot *but* produce loops equal in size and uniformly distributed.

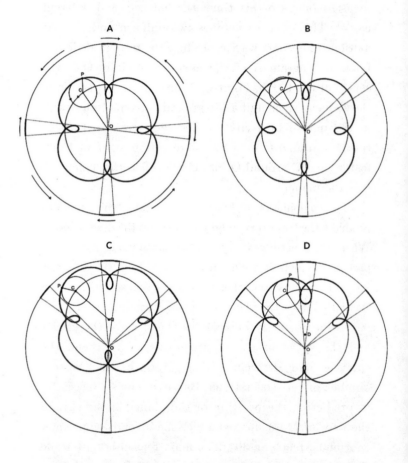

The epicycle and deferent system (A and B), the eccentric model (C) and the Ptolemaic equant model (D).

Imagine that the real loops of a planet are represented by the slim sectors between the dotted lines of diagram B. The places where the loops should take place are not uniformly distributed (the loops are closer in the upper part), nor do they have the same size (the loop of the bottom is smaller than that of the top). Two modifications must therefore be introduced in the model in order to solve the two problems: to locate the loops at the correct place and to make them the correct size. We know that some astronomers, probably Hipparchus, tried to solve the first problem, but before Ptolemy nobody had been able to solve the second one.

The Eccentric Model

In order to locate the loops at the correct place, one can move the centre of the deferent towards the direction in which the loops are closer; see diagram C. Keeping the Earth and the observer at O, we move the entire system up, so that the centre of the deferent is now Q. This model is called an eccentric model because we have made the deferent non-concentric with respect to the centre of the universe, which, of course, remains at O. From Q, the four loops are equally distributed and equal in size, but, from the point of view of the observer at O the loops are non-uniformly distributed. We are thus able to locate them at the correct places.

However, this move has a high cost. Compare how the upper and lower loops of diagrams B and C fit with the limits set by the dotted lines. Now the apparent sizes of the loops

27

are even worse than before. This implies an error of up to 15 ° in the case of Mars. This issue seemed to have no solution: in attempting to solve one of the problems, we have made the other worse. Hipparchus did not propose a solution. For that we must wait for a couple of centuries, to the time of Ptolemy.

The Introduction of the Equant Point

In his *Almagest*, Ptolemy introduces a revolutionary modification on the eccentric model; see diagram D. In his model, C moves uniformly from Q, but the centre of the deferent is D. Ptolemy proposed for the first time that the motion is uniform when measured not from the center of the orbit (D) but from another point (Q), which he called 'equant'. Ptolemy finds that the eccentric point D bisects the equant point, i.e., that QD = DO. The introduction of the equant changes the sizes of the loops, leaving them at the place where the eccentric model placed it.

This significantly improves the accuracy of the model, as is evident when comparing the loops with the dotted lines. The remaining error is always smaller than 1° in the case of Mars, the most complicated planet. Such accuracy would not be superseded until the new observations of Tycho Brahe (1546–1601 CE), the great observer at the end of the sixteenth century, put them in evidence.

The Conqueror of the World

The influence of the *Almagest* in the history of astronomy can hardly be exaggerated. It was the canonical text of astronomy from Ptolemy until Copernicus (1473–1543 CE) and beyond. As the great historian of astronomy Noel Swerdlow observed: 'Ptolemy's model was the best approximation to Kepler's laws before Kepler's laws'. The same Tycho Brahe who contributed through his accurate observations to the revolution that Kepler finally produced also wrote a moving poem dedicated to Ptolemy on 1 October 1584.

By comparing the military conquests of Rome with the scientific conquests of Ptolemy, the poem shows the enormous prestige that the Alexandrian astronomer still had, 15 centuries after his death. It reads:

Go, now, Rome! Exalt your triumphs over the Nile! But this one, born close to the Nile, was more victorious than you. You conquered a part of the world by spilling a lot of blood. He alone conquered the entire world, without any blood. You never dared to rise to the heavens, but he subjected all the stars to his genius. And what is most surprising: your glory and your fame were buried; the one that could defeat others, defeated lies. He remains forever in possession of the sky and preserves an eternal name in the world.

It seems that Ptolemy was right when he wrote his own poem: by studying the stars that rotate without ceasing, he achieved immortality.

Christián C. Carman

Claudius Ptolemy

The Almagest

Translated by G.J. Toomer

Abridged by Christián C. Carman

Book I

1. Preface[1]

T he true philosophers, Syrus,[2] were, I think, quite right to distinguish the theoretical part of philosophy from the practical. Fm: even if practical philosophy, before it is practical, turns out to be theoretical,[3] nevertheless one can see that there is a great

---

1   This 'philosophical' preface and its relationship to Ptolemy's attitude to philosophy is discussed by Boll; *Studien* 68–76, to which the reader is referred for the relevant passages in ancient literature. The general standpoint is Aristotelian.

2   Syrus is also the addressee of a number of other works by Ptolemy (see Toomer[5] 187). Nothing is known about him. The name is very common in (but not confined to) Greco-Roman Egypt. The statement in a scholion to the Tetrabiblos (quoted by Boll, *Studien* 67, n. 2) that some say he was a fictitious person, others that he was a doctor, merely reveals that he was equally unknown in late antiquity.

3   Théon in his commentary (Rome II 320, 13–14) gives φησί...συμβεβηκέναι τῷ πρακτικῷ τό πρότερον αὐτοῦ τοῦ θεωρητικοῦ τυγχάνειν. This is a paraphrase rather than a different reading, but shows that he understood the text as I have translated it. By this obscure expression I take Ptolemy to mean that before actually practising virtues one must have some concept of them (even though this is innate rather than taught).

difference between the two: in the first place, it is possible for many people to possess some of the moral virtues even without being taught, whereas it is impossible to achieve theoretical understanding of the universe without instruction; furthermore, one derives most benefit in the first case [practical philosophy] from continuous practice in actual affairs, but in the other [theoretical philosophy] from making progress in the theory. Hence we thought it fitting to guide our actions (under the impulse of our actual ideas [of what is to be done]) in such away as never to forget, even in ordinary affairs, to strive for a noble and disciplined disposition, but to devote most of our time to intellectual matters, in order to teach theories, which are so many and beautiful, and especially those to which the epithet 'mathematical' is particularly applied. For Aristotle divides theoretical philosophy too, very fittingly, into three primary categories, physics, mathematics and theology.[4] For everything that exists is composed of matter, form and motion; none of these [three] can be observed in its substratum by itself, without the others: they can only be imagined. Now the first cause of the first motion of the universe, if one considers it simply, can be thought of as an invisible and motionless deity; the division [of theoretical philosophy] concerned with investigating this [can be called] 'theology', since

---

4   E.g. Metaphysics E I, IO26a 18 ff., ὥστε τρεῖς ἀν εἶν φιλοσοφίαι θεωρητικαί, μαθηματική, φνσική, θεολογική.

this kind of activity, somewhere up in the highest reaches of the universe, can only be imagined, and is completely separated from perceptible reality. The division [of theoretical philosophy] which investigates material and ever-moving nature, and which concerns itself with 'white', 'hot', 'sweet', 'soft' and suchlike qualities one may call 'physics'; such an order of being is situated (for the most part) amongst corruptible bodies and below the lunar sphere. That division [of theoretical philosophy] which determines the nature involved in forms and motion from place to place, and which serves to investigate shape, number, size, and place, time and suchlike, one may define as 'mathematics'. Its subject-matter falls as it were in the middle between the other two, since, firstly, it can be conceived of both with and without the aid of the senses, and, secondly, it is an attribute of all existing things without exception, both mortal and immortal: for those things which are perpetually changing in their inseparable form, it changes with them, while for eternal things which have an aethereal[5] nature, it keeps their unchanging form unchanged.

From all this we concluded:[6] that the first two divisions of theoretical philosophy should rather be

---

5   'aethereal' (αἰθερώδης) has a precise meaning in Aristotelian physics: everything above the sphere of the moon is composed of an 'incorruptible' substance, unlike anything-known on earth in its consistency (very thin) and in its natural motion (circular). See I 3. One of the names for this substance is 'aether', another 'fifth essence'. See Campanus IV n. 56, pp. 394–5.

6   In this exaltation of mathematics above the other two divisions of philosophy Ptolemy parts company with Aristotle, for whom theology was the most noble pursuit for the human mind.

called guesswork than knowledge, theology because of
its completely invisible and ungraspable nature, physics
because of the unstable and unclear nature of matter;
hence there is no hope that philosophers will ever be
agreed about them; and that only mathematics can
provide sure and unshakeable knowledge to its devotees,
provided one approaches it rigorously. For its kind
of proof proceeds by indisputable methods, namely
arithmetic and geometry. Hence we were drawn to the
investigation of that part of theoretical philosophy, as
far as we were able to the whole of it, but especially to
the theory concerning divine and heavenly things. For
that alone is devoted to the investigation of the eternally
unchanging. For that reason it too can be eternal and
unchanging (which is a proper attribute of knowledge) in
its own domain, which is neither unclear nor disorderly.
Furthermore it can work in the domains of the other
[two divisions of theoretical philosophy] no less than
they do. For this is the best science to help theology
along its way, since it is the only one which can make
a good guess at [the nature of] that activity which is
unmoved and separated; [it can do this because] it is
familiar with the attributes of those beings[7] which are
on the one hand perceptible, moving and being moved,
but on the other hand eternal and unchanging, [I mean
the attributes] having to do with motions and the

7   The heavenly bodies.

34

arrangements of motions. As for physics, mathematics can make a significant contribution. For almost every peculiar attribute of material nature becomes apparent from the peculiarities of its motion from place to place. [Thus one can distinguish] the corruptible from the incorruptible by [whether it undergoes] motion in a straight line or in a circle, and heavy from light, and passive from active, by [whether it moves] towards the centre or away from the centre. With regard to virtuous conduct in practical actions and character, this science, above all things, could make men see clearly; from the constancy, order, symmetry and calm which are associated with the divine, it makes its followers lovers of this divine beauty, accustoming them and reforming their natures, as it were, to a similar spiritual state.

It is this love of the contemplation of the eternal and unchanging which we constantly strive to increase, by studying those parts of these sciences which have already been mastered by those who approached them in a genuine spirit of enquiry, and by ourselves attempting to contribute as much advancement as has been made possible by the additional time between those people and ourselves.[8] We shall try to note down[9] everything which

---

8   This notion of the advancement of science, and particularly astronomy, by the additional time available is one to which Ptolemy recurs in the epilogue, and also, in a specifically astronomical context, at VII 1 and VII 3.

9   ὑπομνηματίσασθι. A ὑπόμνημα is a 'memoir', usually implying summary brevity. Ptolemy recurs to this too in the epilogue.

we think we have discovered up to the present time; we shall do this as concisely as possible and in a manner which can be followed by those who have already made some progress in the field.[10] For the sake of completeness in our treatment we shall set out everything useful for the theory of the heavens in the proper order, but to avoid undue length we shall merely recount what has been adequately established by the ancients. However, those topics which have not been dealt with [by our predecessors] at all, or not as usefully as they might have been, will be discussed at length, to the best of our ability.

2. On the order of the theorems

In the treatise which we propose, then, the first order of business is to grasp the relationship of the earth taken as a whole to the heavens taken as a whole. In the treatment of the individual aspects which follows, we must first discuss the position of the ecliptic[11] and the regions of our part of the inhabited world and also the features differentiating each from the others due to the [varying] latitude at each horizon taken in order.[12] For if the theory of these matters is treated first it will make examination of the rest easier. Secondly, we have to go through the motion of the sun

---

10   Ptolemy assumes that his readers will have a certain competence.

11   I 12-16. The mathematical section I 10-11 is not specifically mentioned here.

12   Book II.

and of the moon, and the phenomena accompanying these [motions];[13] for it would be impossible to examine the theory of the stars[14] thoroughly without first having a grasp of these matters. Our final task in this way of approach is the theory of the stars. Here too it would be appropriate to deal first with the sphere of the so-called 'fixed stars',[15] and follow that by treating the five 'planets', as they are called.[16] We shall try to provide proofs in all of these topics by using as starting-points and foundations, as it were, for our search the obvious phenomena, and those observations made by the ancients and in our own times which are reliable. We shall attach the subsequent structure of ideas to this [foundation] by means of proofs using geometrical methods.

The general preliminary discussion covers the following topics: the heaven is spherical in shape, and moves as a sphere; the earth too is sensibly spherical in shape, when taken as a whole; in position it lies in the middle of the heavens very much like its centre; in size and distance it has the ratio of a point to the sphere of the fixed stars; and it has no motion from place to place. We shall briefly discuss each of these points for the sake of reminder.

---

13   Books III–VI.

14   'Stars' here and throughout chs. 3–8 includes both fixed stars and planets and also, sometimes, sun and moon.

15   Books VII–VIII.

16   Books IX–XIII.

3. That the heavens move like a sphere[17]

It is plausible to suppose that the ancients got their first notions on these topics from the following kind of observations. They saw that the sun, moon and other stars were carried from east to west along circles which were always parallel to each other, that they began to rise up from below the earth itself, as it were, gradually got up high, then kept on going round in similar fashion and getting lower, until, falling to earth, so to speak, they vanished completely, then, after remaining invisible for some time, again rose afresh and set; and [they saw] that the periods of these [motions], and also the places of rising and setting, were, on the whole, fixed and the same.

What chiefly led them to the concept of a sphere was the revolution of the ever-visible stars, which was observed to be circular, and always taking place about one centre, the same [for all]. For by necessity that point became [for them] the pole of the heavenly sphere: those stars which were closer to it revolved on smaller circles, those that were farther away described circles ever greater in proportion to their distance, until one reaches the distance of the stars which become invisible. In the case of these, too, they saw that those near the ever-visible stars remained invisible for a short time, while those farther away remained invisible for a long time, again in proportion [to their distance]. The result was that in the

---

17   See Pedersen 36–7.

beginning they got to the aforementioned notion solely from such considerations; but from then on, in their subsequent investigation, they found that everything else accorded with it, since absolutely all phenomena are in contradiction to the alternative notions which have been propounded.

For if one were to suppose that the stars' motion takes place in a straight line towards infinity, as some people have thought,[18] what device could one conceive of which would cause each of them to appear to begin their motion from the same starting-point every day? How could the stars turn back if their motion is towards infinity? Of, if they did turn back, how could this not be obvious? [On such a hypothesis], they must gradually diminish in size until they disappear, whereas, on the contrary, they are seen to be greater at the very moment of their disappearance, at which time they are gradually obstructed and cut off, as it were, by the earth's surface.

But to suppose that they are kindled as they rise out of the earth and are extinguished again as they fall to earth is a completely absurd hypothesis.[19] For even if we were to concede that the strict order in their size and number, their intervals, positions and periods could

18   According to Théon's commentary (Rome II 338) this belief was Epicurean, but I know of no other evidence. The only other relevant passage appears to be Xenophanes, Diels-Kranz A41a (the sun really moves towards infinity).

19   Théon (Rome II 340) ascribes this to Heraclitus. Otherwise it is attested for Xenophanes (Diels Kranz A38), and was admitted as one possible explanation by Epicurus (e.g. *Letter to Pythocles* 92) and his followers.

be restored by such a random and chance process; that one whole area of the earth has a kindling nature, and another an extinguishing one, or rather that the same part [of the earth] kindles for one set of observers and extinguishes for another set; and that the same stars are already kindled or extinguished for some observers while they are not yet for others: even if, I say, we were to concede all these ridiculous consequences, what could we say about the ever-visible stars, which neither rise nor set? Those stars which are kindled and extinguished ought to rise and set for observers everywhere, while those which are not kindled and extinguished ought always to be visible for observers everywhere. What cause could we assign for the fact that this is not so? We will surely not say that stars which are kindled and extinguished for some observers never undergo this process for other observers. Yet it is utterly obvious that the same stars rise and set in certain regions [of the earth] and do neither at others.

To sum up, if one assumes any motion whatever, except spherical, for the heavenly bodies, it necessarily follows that their distances, measured from the earth upwards, must vary, wherever and however one supposes the earth itself to be situated. Hence the sizes and mutual distances of the stars must appear to vary for the same observers during the course of each revolution, since at one time they must be at a greater distance, at another at a lesser. Yet we see that no such variation occurs.

For the apparent increase in their sizes at the horizons[20] is caused, not by a decrease in their distances, but by the exhalations of moisture surrounding the earth being interposed between the place from which we observe and the heavenly bodies, just as objects placed in water appear bigger than they are, and the lower they sink, the bigger they appear.

The following considerations also lead us to the concept of the sphericity of the heavens. No other hypothesis but this can explain how sundial constructions produce correct results; furthermore, the motion of the heavenly bodies is the most unhampered and free of all motions, and freest motion belongs among plane figures to the circle and among solid shapes to the sphere; similarly, since of different shapes having an equal boundary those with more angles are greater [in area or volume], the circle is greater than [all other] surfaces, and the sphere greater than [all other] solids;[21] [likewise] the heavens are greater than all other bodies.

Furthermore, one can reach this kind of notion from certain physical considerations. E.g., the aether is, of all bodies, the one with constituent parts which are finest

20 Ptolemy refers to the well-known phenomenon that the sun and moon appear larger when close to the horizon. He gives an incorrect physical and optical explanation here. In a later work (*Optics* III 60, ed. Lejeune p. 116) he correctly explains it as a purely psychological phenomenon. No doubt instrumental measurement of the apparent diameters had convinced him that the enlargement is entirely illusory.

21 These propositions were proved in a work by Zenodorus (early second century B.C., see Toomer[I]) from which extensive excerpts are given by (among others) Théon (Rome II 355–79).

and most like each other; now bodies with parts like each other have surfaces with parts like each other; but the only surfaces with parts like each other are the circular, among planes, and the spherical, among three-dimensional surfaces. And since the aether is not plane, but three-dimensional, it follows that it is spherical in shape. Similarly, nature formed all earthly and corruptible bodies out of shapes which are round but of unlike parts, but all aethereal and divine bodies out of shapes which are of like parts and spherical. For if they were flat or shaped like a discus[22] they would not always display a circular shape to all those observing them simultaneously from different places on earth. For this reason it is plausible that the aether surrounding them, too, being of the same nature, is spherical, and because of the likeness of its parts moves in a circular and uniform fashion.

4. That the Earth too, taken as a whole, is sensibly spherical[23]

That the earth, too, taken as a whole,[24] is sensibly spherical can best be grasped from the following considerations. We can see, again, that the sun, moon and other stars do not rise and set simultaneously for everyone on earth, but

---

22 The only relevant passage I know is Empedocles, Diels-Kranz A60, who maintained that the moon is disk-shaped.

23 See Pedersen 37–9.

24 'taken as a whole': ignoring local irregularities such as mountains, which are negligible in comparison to the total mass.

do so earlier for those more towards the east, later for those towards the west. For we find that the phenomena at eclipses, especially lunar eclipses,[25] which take place at the same time [for all observers], are nevertheless not recorded as occurring at the same hour (that is at an equal distance from noon) by all observers. Rather, the hour recorded by the more easterly observers is always later than that recorded by the more westerly. We find that the differences in the hour are proportional to the distances between the places [of observation]. Hence one can reasonably conclude that the earth's surface is spherical, because its evenly curving surface (for so it is when considered as a whole) cuts off [the heavenly bodies] for each set of observers in turn in a regular fashion. If the earth's shape were any other, this would not happen, as one can see from the following arguments. If it were concave, the stars would be seen rising first by those more towards the west; if it were plane, they would rise and set simultaneously for everyone on earth; if it were triangular or square or any other polygonal shape, by a similar argument, they would rise and set simultaneously for all those living on the same plane surface. Yet it is apparent that nothing like this takes place. Nor could it be cylindrical, with the curved surface in the east-west direction, and the flat sides towards the poles of the universe, which some might suppose more

---

25   The timings for solar eclipses are complicated by parallax.

plausible. This is clear from the following: for those living on the curved surface none of the stars would be ever-visible, but either all stars would rise and set for all observers, or the same stars, for an equal [celestial] distance from each of the poles, would always be invisible for all observers. In fact, the further we travel toward the north, the more[26] of the southern stars disappear and the more of the northern stars appear. Hence it is clear that here too the curvature of the earth cuts off [the heavenly bodies] in a regular fashion in a north-south direction, and proves the sphericity [of the earth] in all directions.

There is the further consideration that if we sail towards mountains or elevated places from and to any direction whatever, they are observed to increase gradually in size as if rising up from the sea itself in which they had previously been submerged: this is due to the curvature of the surface of the water.

5. That the Earth is in the middle of the heavens[27]

Once one has grasped this, if one next considers the position of the earth, one will find that the phenomena associated with it could take place only if we assume that it is in the middle of the heavens, like the centre of

---

26   Reading πλείονα (with D) for τά πλείονα. Corrected by Manitius.

27   See Pedersen 39–42..

a sphere. For if this were not the case, the earth would have to be either:

1. not on the axis [of the universe] but equidistant from both poles, or
2. on the axis but removed towards one of the poles, or
3. neither on the axis nor equidistant from both poles.

Against the first of these three positions militate the following arguments. If we imagined [the earth] removed towards the zenith or the nadir of some observer, then, if he were at *sphaera recta*, he would never experience equinox, since the horizon would always divide the heavens into two unequal parts, one above and one below the earth; if he were at *sphaera obliqua*, either, again, equinox would never occur at all, or, [if it did occur,] it would not be at a position halfway between summer and winter solstices, since these intervals would necessarily be unequal, because the equator, which is the greatest of all parallel circles drawn about the poles of the [daily] motion, would no longer be bisected by the horizon; instead [the horizon would bisect] one of the circles parallel to the equator, either to the north or to the south of it. Yet absolutely everyone agrees that these intervals are equal everywhere on earth, since [everywhere] the increment of the longest day over the equinoctial day at the summer solstice is equal to the decrement of the shortest day from the equinoctial day at the winter solstice. But if, on the other hand, we imagined the displacement to be towards the east or west of some observer, he would find that the sizes and distances of the stars would not remain constant and unchanged at eastern and western horizons, and that the time-interval from rising to culmination would not be equal to the interval from culmination to setting. This is obviously completely in disaccord with the phenomena.

Against the second position, in which the earth is imagined to lie on the axis removed towards one of the poles, one can make the following objections. If this were so, the plane of the horizon would divide the heavens into a part above the earth and a part below the earth which are unequal and always different for different latitudes,[28] whether one considers the relationship of the same part at two different latitudes or the two parts at the same latitude. Only at *sphaera recta* could the horizon bisect the sphere; at a *sphaera obliqua* situation such that the nearer pole were the ever-visible one, the horizon would always make the part above the earth lesser and the part below the earth greater; hence another phenomenon would be that the great circle of the ecliptic would be divided into unequal parts by the plane of the horizon. Yet it is apparent that this is by no means so. Instead, six zodiacal signs are visible above the earth at all times and places, while the remaining six are invisible; then again [at a later time] the latter are visible in their entirety above the earth, while at the same time the others are not visible. Hence it is obvious that the horizon

---

28   The word translated here and elsewhere as '[terrestrial] latitude' is κλίμα, for the meaning of which see Introduction.

bisects the zodiac, since the same semi-circles are cut off by it, so as to appear at one time completely above the earth, and at another [completely] below it.

And in general, if the earth were not situated exactly below the [celestial] equator, but were removed towards the north or south in the direction of one of the poles, the result would be that at the equinoxes the shadow of the gnomon at sunrise would no longer form a straight line with its shadow at sunset in a plane parallel to the horizon, not even sensibly.[29] Yet this is a phenomenon which is plainly observed everywhere.

It is immediately clear that the third position enumerated is likewise impossible, since the sorts of objection which we made to the first [two] will both arise in that case.

To sum up, if the earth did not lie in the middle [of the universe], the whole order of things which we observe in the increase and decrease of the length of daylight would be fundamentally upset. Furthermore, eclipses of the moon would not be restricted to situations where the moon is diametrically opposite the sun (whatever part of the heaven [the luminaries are in]), since the earth would often come between them when they were not diametrically opposite, but at intervals of less than a semi-circle.

---

29   The *caveat* 'sensibly' is inserted because the equinox is not a date but an instant of time. Therefore on the day of equinox the sun does not rise due east and set due west (as is implied by the rising and setting shadows lying on the same straight line). However, the difference would be 'imperceptible to the senses'.

6. That the Earth has the ratio of a point to the heavens[30]

Moreover, the earth has, to the senses, the ratio of a point to the distance of the sphere of the so-called fixed stars.[31] A strong indication of this is the fact that the sizes and distances of the stars, at any given time, appear equal and the same from all parts of the earth everywhere, as observations of the same [celestial] objects from different latitudes are found to have not the least discrepancy from each other. One must also consider the fact that gnomons set up in any part of the earth whatever, and likewise the centres of armillary spheres,[32] operate like the real centre of the earth; that is, the lines of sight [to heavenly bodies] and the paths of shadows caused by them agree as closely with the [mathematical] hypotheses explaining the phenomena as if they actually passed through the real centre-point of the earth.

Another clear indication that this is so is that the planes drawn through the observer's lines of sight at any point [on earth], which we call 'horizons', always bisect the whole heavenly sphere. This would not happen if the earth were of perceptible size in relation to the distance of the heavenly

---

30 See Pedersen 42–3.

31 Ptolemy qualifies the traditional terminology for the fixed stars as 'so-called' (καλουμένων) because they do in fact, according to him, have a motion (the modern 'precession'). He develops the point further at VII 1. In general, however, he uses the traditional terminology without qualification.

32 An example of an armillary sphere (κρκιωτη σφαίρα) is the 'astrolabe' described in VI. For references to the term in other works see LSJ s.v. κρικωτός.

bodies; in that case only the plane drawn through the centre of the earth could bisect the sphere, while a plane through any point on the surface of the earth would always make the section [of the heavens] below the earth greater than the section above it.

7. That the Earth does not have any motion from place to place, either[33]

One can show by the same arguments as the preceding that the earth cannot have any motion in the aforementioned directions, or indeed ever move at all from its position at the centre. For the same phenomena would result as would if it had any position other than the central one. Hence I think it is idle to seek for causes for the motion of objects towards the centre, once it has been so clearly established from the actual phenomena that the earth occupies the middle place in the universe, and that all heavy objects are carried towards the earth. The following fact alone would most readily lead one to this notion [that all objects fall towards the centre]. In absolutely all parts of the earth, which, as we said, has been shown to be spherical and in the middle of the universe, the direction[34] and path of the motion (I mean

---

33 See Pedersen 43–4.

34 πρόσνευσις which I have translated 'the direction of motion' here, means basically 'direction in which something points' (for astronomical usages see V 5 and VI II). Thus it would also include here the direction of a plumb-line.

the proper, [natural] motion) of all bodies possessing weight is always and everywhere at right angles to the rigid plane drawn tangent to the point of impact. It is clear from this fact that, if [these falling objects] were not arrested by the surface of the earth, they would certainly reach the centre of the earth itself, since the straight line to the centre is also always at right angles to the plane tangent to the sphere at the point of intersection [of that radius] and the tangent.

Those who think it paradoxical that the earth, having such a great weight, is not supported by anything and yet does not move, seem to me to be making the mistake of judging on the basis of their own experience instead of taking into account of the peculiar nature of the universe. They would not, I think, consider such a thing strange once they realised that this great bulk of the earth, when compared with the whole surrounding mass [of the universe], has the ratio of a point to it. For when one looks at it in that way, it will seem quite possible that that which is relatively smallest should be overpowered and pressed in equally from all directions to a position of equilibrium by that which is the greatest of all and of uniform nature. For there is no up and down in the universe with respect to itself,[35] any more than one could imagine such a thing in a sphere: instead the proper and natural motion of the compound

---

35   Reading αὐτόν (with D, Is) for αὐτήν at H23,1

bodies in it is as follows: light and rarefied bodies drift outwards towards the circumference, but seem to move in the direction which is 'up' for each observer, since the overhead direction for all of us, which is also called 'up', points towards the surrounding surface;[36] heavy and dense bodies, on the other hand, are carried towards the middle and the centre, but seem to fall downwards, because, again, the direction which is for all us towards our feet, called 'down', also points towards the centre of the earth. These heavy bodies, as one would expect, settle about the centre because of their mutual pressure and resistance, which is equal and uniform from all directions. Hence, too, one can see that it is plausible that the earth, since its total mass is so great compared with the bodies which fall towards it, can remain motionless under the impact of these very small weights (for they strike it from all sides), and receive, as it were, the objects falling on it. If the earth had a single motion in common with other heavy objects, it is obvious that it would be carried down faster than all of them because of its much greater size: living things and individual heavy objects would be left behind, riding on the air, and the earth itself would very soon have fallen completely out of the heavens. But such things are utterly ridiculous merely to think of.

---

36   It is not clear to me whether Ptolemy means the outmost boundary of the universe or merely the surface (of the 'aether') surrounding the *earth*.

But certain people,[37] [propounding] what they consider a more persuasive view, agree with the above, since they have no argument to bring against it, but think that there could be no evidence to oppose their view if, for instance, they supposed the heavens to remain motionless, and the earth to revolve from west to east about the same axis [as the heavens], making approximately one revolution each day;[38] or if they made both heaven and earth move by any amount whatever, provided, as we said, it is about the same axis, and in such a way as to preserve the overtaking of one by the other. However, they do not realise that, although there is perhaps nothing in the celestial phenomena which would count against that hypothesis, at least from simpler considerations, nevertheless from what would occur here on earth and in the air, one can see that such a notion is quite ridiculous. Let us concede to them [for the sake of argument] that such an unnatural thing could happen as that the most rare and light of matter should either not move at all or should move in a way no different from that of matter with the opposite nature (although things in the air, which are less rare [than the heavens] so obviously move with a more rapid motion than any earthy object); [let us concede that] the densest

---

37   Heraclides of Pontos (late fourth century B.C.) is the earliest certain authority for the view that the earth rotates on its axis. See *HAMA* II 694–6. It was also adopted by Aristarchus as part of his more radical heliocentric hypothesis.

38   'approximately' because one revolution takes place in a sidereal, not a solar day.

and heaviest objects have a proper motion of the quick and uniform kind which they suppose (although, again, as all agree, earthy objects are sometimes not readily moved even by an external force). Nevertheless, they would have to admit that the revolving motion of the earth must be the most violent of all motions associated with it, seeing that it makes one revolution in such a short time; the result would be that all objects not actually standing on the earth would appear to have the same motion, opposite to that of the earth: neither clouds nor other flying or thrown objects would ever be seen moving towards the east, since the earth's motion towards the east would always outrun and overtake them, so that all other objects would seem to move in the direction of the west and the rear. But if they said that the air is carried around in the same direction and with the same speed as the earth, the compound objects in the air would none the less always seem to be left behind by the motion of both [earth and air]; or if those objects too were carried around, fused, as it were, to the air, then they would never appear to have any motion either in advance or rearwards: they would always appear still, neither wandering about nor changing position, whether they were flying or thrown objects. Yet we quite plainly see that they do undergo all these kinds of motion, in such a way that they are not even slowed down or speeded up at all by any motion of the earth.

8. That there are two different primary motions in the heavens[39]

It was necessary to treat the above hypotheses first as an introduction to the discussion of particular topics and what follows after. The above summary outline of them will suffice, since they will be completely confirmed and further proven by the agreement with the phenomena of the theories which we shall demonstrate in the following sections. In addition to these hypotheses, it is proper, as a further preliminary, to introduce the following general notion, that there are two different primary motions in the heavens. One of them is that which carries everything from east to west: it rotates them with an unchanging and uniform motion along circles parallel to each other, described, as is obvious, about the poles of this sphere which rotates everything uniformly. The greatest of these circles is called the 'equator',[40] because it is the only [such parallel circle] which is always bisected by the horizon (which is a great circle), and because the revolution which the sun makes when located on it produces equinox everywhere, to the senses. The other motion is that by which the spheres of the stars perform movements in the opposite sense to the first motion, about another pair of poles, which are

---

39   See Pedersen 45.

40   'equator': ἰσημερινός, literally 'of equal day' or 'equinoctial'. See Introduction.

different from those of the first rotation. We suppose that this is so because of the following considerations. When we observe for the space of any given single day, all heavenly objects whatever are seen, as far as the senses can determine, to rise, culminate and set at places which are analogous and lie on circles parallel to the equator; this is characteristic of the first motion. But when we observe continuously without interruption over an interval of time, it is apparent that while the other stars retain their mutual distances and (for a long time) the particular characteristics arising from the positions they occupy as a result of the first motion,[41] the sun, the moon and the planets have certain special motions which are indeed complicated and different from each other, but are all, to characterise their general direction,[42] towards the east and opposite to [the motion of] those stars which preserve their mutual distances and are, as it were, revolving on one sphere.

Now if this motion of the planets too took place along circles parallel to the equator, that is, about the poles which produce the first kind of revolution, it would be sufficient to assign a single kind of revolution to all alike, analogous to the first. For in that case it would have seemed plausible that the movements which they undergo are caused by various retardations, and not

---

41   These characteristics of the fixed stars are e.g. dates of first and last visibility.
They are unchanged 'for a long time' because the effect of precession is very slow.

42   The qualification is inserted here to allow for the retrogradations of the planets.

by a motion in the opposite direction. But as it is, in addition to their movement towards the east, they are seen to deviate continuously to the north and south [of the equator]. Moreover the amount of this deviation cannot be explained as the result of a uniformly-acting force pushing them to the side: from that point of view it is irregular, but it is regular if considered as the result of [motion on] a circle inclined to the equator. Hence we get the concept of such a circle, which is one and the same for all planets, and particular to them. It is precisely defined and, so to speak, drawn by the motion of the sun, but it is also travelled by the moon and the planets, which always move in its vicinity, and do not randomly pass outside a zone on either side of it which is determined for each body. Now since this too is shown to be a great circle, since the sun goes to the north and south of the equator by an equal amount, and since, as we said, the eastward motion of ail of the planets takes place on one and the same circle, it became necessary to suppose that this second, different motion of the whole takes place about the poles of the inclined circle we have defined [i.e. the ecliptic], in the opposite direction to the first motion.

If, then, we imagine a great circle drawn through the poles of both the above mentioned circles, (which will necessarily bisect each of them, that is the equator and the circle inclined to it [the ecliptic], at right angles), we will have four points on the ecliptic: two will be produced by [the intersection of] the equator, diametrically opposite each other; these are called 'equinoctial' points. The one at which the motion [of the planets] is from south to north is called the 'spring' equinox, the other the 'autumnal'. Two [other points] will be produced by [the intersection of] the circle drawn through both poles; these too, obviously, will be diametrically opposite each other; they are called 'tropical' [or 'solsticial'] points. The one south of the equator is called the 'winter' [solstice], the one north, the 'summer' [solstice].

We can imagine the first primary motion, which encompasses all the other motions, as described and as it were defined by the great circle drawn through both poles [of equator and ecliptic] revolving, and carrying everything else with it, from east to west about the poles of the equator. These poles are fixed, so to speak, on the 'meridian' circle', which differs from the aforementioned [great] circle in the single respect that it is not drawn through the poles of the ecliptic too at all positions of the latter. Moreover, it is called 'meridian' because it is considered to be always orthogonal to the horizon. For a circle in such a position divides both hemispheres, that above the earth and that below it, into two equal parts, and defines the midpoint of both day and night.

The second, multiple-part motion is encompassed by the first and encompasses the spheres of all the planets. As we said, it is carried around by the aforementioned [first motion], but itself goes in the opposite direction about the poles of the ecliptic, which are also fixed on the circle which produces the first motion, namely the circle through both poles [of ecliptic and equator]. Naturally they [the poles of the ecliptic] are carried around with it [the circle through both poles], and, throughout the period of the second motion in the opposite direction, they always keep the great circle of the ecliptic, which is described by that [second] motion, in the same position with respect to the equator.[43]

---

43   My translation follows the interpretation of Théon (Rome II 447). Manitius (p. 24 n. a.) wrongly considers τού γραφομένου κάι λοξοῦ κύκλου interpolated, partly because he misinterprets συντηροῦσιν (which is used here in a way similar to συντηροῦσαν at HI 6,10).

Book III

1. Preface[44]

In the preceding part of our treatise we have dealt with those aspects of heaven and earth which required, in outline, a preliminary mathematical discussion; also the inclination of the sun's path through the ecliptic, and the resultant particular phenomena, both at *sphaera recta* and at *sphaera obliqua* for every inhabited region. We think that we should [now] discuss, as the subject which appropriately follows the above, the theory of the sun and moon, and go through the phenomena which are a consequence of their motions. For none of the phenomena associated with the [other] heavenly bodies can be completely investigated without the previous treatment of these [two]. Furthermore,

---

44   D and part of the Arabic tradition (L, P, but not Q, T) begin chapter 1 at this point.

we find that the subject of the sun's motion must take first place amongst these [sun and moon], since without that it would, again, be impossible to give a complete discussion of the moon's theory from start to finish.

3. On the hypotheses for uniform circular motion[45]

Our next task is to demonstrate the apparent anomaly of the sun. But first we must make the general point that the rearward displacements of the planets with respect to the heavens are, in every case, just like the motion of the universe in advance, by nature uniform and circular. That is to say, if we imagine the bodies or their circles being carried around by straight lines, in absolutely every case the straight line in question describes equal angles at the centre of its revolution in equal times. The apparent irregularity [anomaly] in their motions is the result of the position and order of those circles in the sphere of each by means of which they carry out their movements, and in reality there is in essence nothing alien to their eternal nature in the 'disorder' which the phenomena are supposed to exhibit. The reason for the appearance of irregularity can be explained by two hypotheses, which are the most basic and simple. When their motion is viewed with respect to a circle imagined to be in the plane of the ecliptic, the centre of which coincides with the centre of the

---

45   See *HAMA* 55–7, Pedersen 134–44..

universe (thus its centre can be considered to coincide with our point of view), then we can suppose, either that the uniform motion of each [body] takes place on a circle which is not concentric with the universe, or that they have such a concentric circle, but their uniform motion takes place, not actually on that circle, but on another circle, which is carried by the first circle, and [hence] is known as the 'epicycle'. It will be shown that either of these hypotheses will enable [the planets] to appear, to our eyes, to traverse unequal arcs of the ecliptic (which is concentric to the universe) in equal times.

In the eccentric hypothesis: [see Fig. 3.1] we imagine the eccentric circle, on which the body travels with uniform

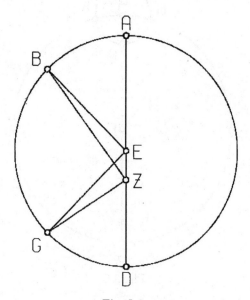

Fig. 3. 1

motion, to be ABGD on centre E, with diameter AED, on which point Z represents the observer.[46] Thus A is the apogee, and D the perigee. We cut off equal arcs AB and DG, and join BE, BZ, GE and GZ. Then it is immediately obvious that the body will traverse the arcs AB and GD in equal times, but will [in so doing] appear to have traversed unequal arcs of a circle drawn on centre Z. For ∠ BEA = ∠ GED. But ∠ BZA < ∠ BEA (or ∠ GED), and ∠ GZD > ∠ GED (or ∠ BEA).

In the epicyclic hypothesis: we imagine [see Fig. 3.2] the circle concentric with the ecliptic as ABGD on centre

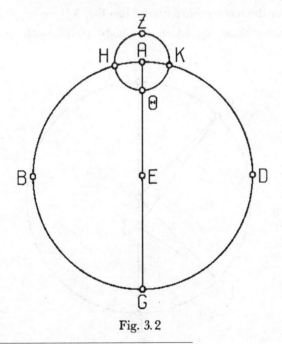

Fig. 3. 2

E, with diameter AEG, and the epicycle carried by it, on which the body moves, as ZHΘK on centre A.

Then here too it is immediately obvious that, as the epicycle traverses circle ABGD with uniform motion, say from A towards B, and as the body traverses the epicycle with uniform motion, then when the body is at points Z and Θ, it will appear to coincide with A, the centre of the epicycle, but when it is at other points it will not. Thus when it is, e.g., at H, its motion will appear greater than the uniform motion [of the epicycle] by arc AH, and similarly when it is at K its motion will appear less than the uniform by arc AK.

Now in this kind of eccentric hypothesis[47] the least speed always occurs at the apogee and the greatest at the perigee, since ∠ AZB [in Fig. 3.1] is always less than ∠ DZG. But in the epicyclic hypothesis both this and the reverse are possible. For the motion of the epicycle is towards the rear with respect to the heavens, say from A towards B [in Fig. 3.2]. Now if the motion of the body on the epicycle is such that it too moves rearwards from the apogee, that is from Z towards H, the greatest speed will occur at the apogee, since at that point both epicycle and body are moving in the same direction. But if the motion of the body from the apogee is in advance on the

---

47 Ptolemy is hinting at the existence of another kind of eccentric hypothesis, one which is geometrically equivalent to that epicyclic hypothesis in which the sense of rotation is the same for both planet and epicycle. But he does not discuss this until XII 1, where we learn that the equivalence was already known to Apollonius of Perge (*c.* 200 B.C.). See *HAMA* 149–50.

epicycle, that is from Z towards K, then the reverse will occur: the least speed will occur at the apogee, since at that point the body is moving in the opposite direction to the epicycle.

Having established that, we must next make the additional preliminary point that for bodies which exhibit a double anomaly both the above hypotheses may be combined, as we shall prove in our discussions of such bodies, but for a body which displays a single invariant anomaly, a single one of the above hypotheses will suffice; and [in this case] all the phenomena will be represented, with no difference, by either hypothesis, provided that the same ratios are preserved in both. By this I mean that the ratio, in the eccentric hypothesis, of the distance between the centre of vision and the centre of the eccentre to the radius of the eccentre, must be the same as the ratio,. in the epicyclic hypothesis, of the radius of the epicycle to the radius of the deferent;[48] and furthermore that the time taken by the body, travelling towards the rear, to traverse the immovable eccentre, must be the same as the time taken by the epicycle, also travelling towards the rear, to traverse the circle with the observer as centre [the deferent], while the body moves with equal [angular] speed about the epicycle, but so that its motion at the apogee [of the epicycle] is in advance. If these conditions are fulfilled, the identical phenomena will result from either

48   'deferent': see Introduction.

hypothesis. We shall briefly show this [now] by comparing the ratios in abstract, and later by means of the actual numbers we shall assign to them for the sun's anomaly.[49] I say then, first, that in both hypotheses, the greatest difference between the uniform motion and the apparent, non-uniform motion (which is also the notional position of the mean speed for the bodies)[50] occurs when the apparent distance from the apogee comprises a quadrant, and that the time between apogee [position] and the above-mentioned mean speed [position] is greater than the time between mean speed and perigee. Hence, for the eccentric hypothesis always, and for the epicyclic hypothesis when the motion at apogee is in advance, the time from least speed to mean is greater than the time from mean speed to greatest; for in both hypotheses the slowest motion takes place at the apogee. But [for the epicyclic hypothesis] when the sense of revolution of the body is rearwards from the apogee on the epicycle, the reverse is true: the time from greatest speed to mean is greater than the time from mean to least, since in this case the greatest speed occurs at the apogee. First, then, [see Fig. 3.3] let the body's eccenter be ABGD on centre E, with diameter AEG. On this diameter take the centre of the ecliptic, that is, the position of the observer, at Z, and draw BZD through Z

---

49   Reference to III 4.

50   Ptolemy never attempts to prove this statement about the position where the apparent motion equals the mean motion, but it is intuitively seen to be true from the epicyclic model. See *HAMA* 57, Pedersen 143.

at right angles to AEG. Let the positions of the body be B and D, so that, obviously, its apparent distance from apogee A is a quadrant on either side. We have to prove that the greatest difference between mean and anomalistic motion takes place at points B and D.

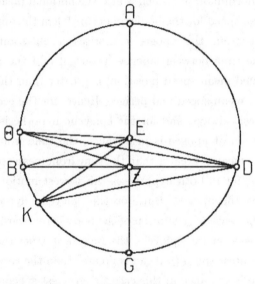

Fig. 3.3

Join EB and ED.

It is immediately obvious that the ratio of ∠ EBZ to 4 right angles equals the ratio of the arc of the difference due to the anomaly[51] to the whole circle; for ∠ AEB subtends the arc of the uniform motion, and ∠ AZB

---

51   This expression is later used as a technical term for the angle corresponding to ∠ EBZ here, and is usually translated 'equation of anomaly'.

subtends the arc of the apparent, non-uniform motion, and the difference between them is ∠ EBZ.

I say, then, that no angle greater than these two [∠ EBZ and ∠ EDZ] can be constructed on line EZ at the circumference of circle ABGD.

[Proof:] Construct at points Θ and K angles EΘZ and EKZ, and join ΘD, KD.

Then since, in any triangle, the greater side subtends the greater angle,[52] and ΘZ > ZD, ∴ ∠ ΘDZ > ∠ DΘZ.

But ∠ EDΘ = ∠ EΘD, since EΘ = ED [radii].

Therefore, by addition, ∠ EDZ (= ∠ EBD) > ∠ EΘZ. Again, since DZ > KZ, ∠ ZKD > ∠ ZDK.

But ∠ EKD = ∠ EDK, since EK = ED.

Therefore, by subtraction, ∠ EDZ (= ∠ EBZ) > ∠ EKZ.

Therefore it is impossible for any other angle to be constructed in the way defined greater than those at points B and D.

Simultaneously it is proven that arc AB, which represents the time from least speed to mean, exceeds BG, which represents the time from mean speed to greatest, by twice the arc comprising the equation of anomaly. For ∠ AEB exceeds a right angle (∠ EZB) by ∠ EBZ, and

___

52 Precisely this statement, that the greater angle is subtended by the greater side, is the enunciation of Euclid I 19 (which Heiberg refers to ad loc.). But in fact what underlies Ptolemy's statement is that, if side *a* is greater than side *b*, angle A is greater than angle B, which is Euclid I 18. Perhaps we should adopt the reading of D, ὑπὸ τὴν μείζονα πλευρὰν ἡ μείζων γωνία ὑποτείνει ('the greater angle subtends the greater side'), and assume that the text has been assimilated to the (wrong) Euclidean wording.

∠ BEG falls short of a right angle by the same amount.

To prove the same theorem again for the other hypothesis, let [Fig. 3.4] the circle concentric with the universe be ABG on centre D and diameter ADB, and let the epicycle which is carried around it in the same plane be EZH on centre A. Let us suppose the body to be at H when its apparent distance from the apogee is a quadrant. Join AH and DHG.

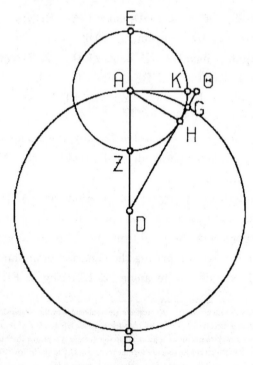

Fig. 3.4

I say that DHG is tangent to the epicycle; for that is the position in which the difference between uniform and anomalistic motion is greatest.

[Proof:] The mean motion, counted from the apogee, is represented by ∠ EAH; for the body traverses the epicycle with the same [angular] speed as the epicycle traverses circle ABG. Furthermore the difference between mean and apparent motion is represented by ∠ ADH. Therefore it is clear that the amount by which ∠ EAH exceeds ∠ ADH (namely ∠ AHD) represents the apparent distance of the body from the apogee. But this distance is, by hypothesis, a quadrant. Therefore ∠ AHD is a right angle, and hence line DHG is tangent to epicycle EZH. Therefore arc AG, since it comprises the distance between the centre A and the tangent, is the greatest possible difference due to the anomaly.

By the same reasoning, arc EH, which according to the sense of rotation on the epicycle assumed here, represents the time from least speed to mean, exceeds arc HZ, which represents the time from mean speed to greatest, by twice arc AG. For if we produce DH to Θ and draw AKΘ at right angles to EZ,

$$\angle \ KAH = \angle \ ADG,^{53}$$
$$\text{and arc } KH = \text{arc } AG.^{54}$$

---

53   Euclid VI 8

54   To get a grammatical text I excise ὁμοία at H225,4. It was introduced (at an early period, since it is reflected in the Arabic translations) as a correction of Ptolemy's inaccurate (to the scholastic mind) statement that arc KH equals arc AG. Since the arcs are on circles of different sizes, they are technically only 'similar'. An alternative correction would be ἴσαι μέν γίγνονται αἵ τε ὑπό KAH καί ΑΔΗ γωνίαι (which is actually found in Théon's commentary

And arc EKH is greater than a quadrant by arc KH,
while arc ZH is less than a quadrant by arc KH.
Q.E.D.

It is also true that the same effects will be produced
by both hypotheses if one takes a partial motion over the
same stretch of time for both, whether one considers the
mean motion or the apparent, or the difference between
them, that is the equation of anomaly. The best way to
see that is as follows.

[See Fig. 3.5.][55] Let the circle concentric with the
ecliptic be ABG on centre D, and let the circle which
is eccentric but equal to the concentre ABG be EZH
on centre Θ. Let the common diameter through their
centres D, Θ and the apogee E be EAΘD. Cut off at
random an arc AB on the concentre, and with centre B
and radius DΘ draw the epicycle KZ. Join KBD.

I say that the body will be carried by both kinds
of motion [i.e. according to both hypotheses] to point
Z, the intersection of the eccentre and the epicycle,
in the same time in all cases (that is, the three arcs,
EZ on the eccentre, AB on the concentre, and KZ on
the epicycle, are all similar), and that the difference
between uniform and anomalistic motion, and the
apparent positions of the body, will turn out to be one

---

ad loc., Rome III 868,8, but is probably a paraphrase; it also seems to be behind L).

55   The figure in Heiberg (p. 225) wrongly omits the letter corresponding to L (though
this is found in all mss.). Manitius, misled by this, 'emended' ΛΛ at H226,23 to the
nonsensical 'AB'.

and the same according to both hypotheses. [Proof:]
Join ZΘ, BZ and DZ.

Since, in the quadrilateral BDΘZ, the opposite
sides are equal, ZΘ to BD and BZ to DΘ, BDΘZ is
a parallelogram.

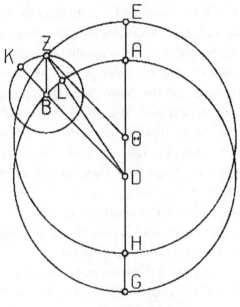

Fig. 3.5

Therefore ∠ EΘZ = ∠ ADB = ∠ ZBK.

Therefore, since they are angles at the centre [of
circles], the arcs subtended by them are also similar, i.e.

Arc EZ of the eccentre ‖ arc AB of the concentre ‖
arc KZ of the epicycle.

Therefore the body will be carried by both kinds of motions in the same time to the same point, Z, and will appear to have traversed the same arc AL of the ecliptic from the apogee, and accordingly the equation of anomaly will be the same in both hypotheses; for we showed that that equation is represented by ∠ DZΘ in the eccentric hypothesis and by ∠ BDZ in the epicyclic hypothesis, and these two angles are alternate and equal, since, as we have shown, ZΘ is parallel to BD.

It is obvious that the same results will hold good for all distances [of the body from the apogee]. For quadrilateral ΘDZB will always be a parallelogram, and [hence] the motion of the body on the epicycle will actually describe the eccentric circle, provided the ratios[56] are similar and their members equal in both hypotheses.

Moreover, even if the members are unequal in size, provided their ratios are similar, the same phenomena will result. This can be shown as follows.

As before [see Fig. 3.6] let the circle concentric with the universe be ABG on centre D and the diameter, on which the body reaches apogee and perigee positions, ADG. Let the epicycle be drawn on point B, at an arbitrary distance, arc AB, from apogee A. Let the arc traversed by the body [on the epicycle] be EZ, which is, obviously, similar to AB, since the revolutions on [both] circles have the same period. Join DBE, BZ, DZ.

---

56   The ratios are e:R and r:R.

Now it is immediately obvious that, according to this [epicyclic] hypothesis, ∠ ADE will always equal ∠ ZBE, and the body will appear to lie on line DZ. But I say that the body will also appear to lie on the same line DZ according to the eccentric hypothesis, whether the eccentre is greater or smaller than the concentre ABG, provided only that one assumes that the ratios are similar and that the periods of revolution are the same.

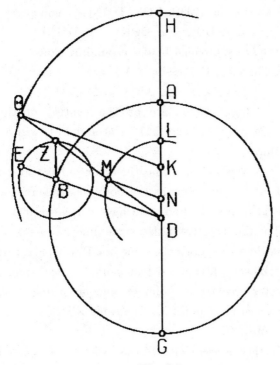

Fig. 3.6

[Proof:] Let the eccentre be drawn under the conditions we have described, greater (than the concentre) as HΘ on centre K ([which must lie] on AG), and smaller [than the concentre] as LM on centre N (this too [must lie on AG]). Produce DZ as DMZΘ, and DA as DLAH, and join ΘK, MN.

Then since

$$DB:BZ = \Theta K:KD = MN:ND \text{ (by hypothesis)},$$

and ∠ BZD = ∠ MDN (since DA is parallel to BZ); the three triangles (ZDB, DΘK, DMN] are equiangular,

$$\text{and } \angle BDZ = \angle D\Theta K = \angle DMN$$

(angles subtended by corresponding sides).

Therefore DB, ΘK and MN are parallel.

$$\therefore \angle ADB = \angle AK\Theta = \angle ANM.$$

Since these angles are at the centres of their circles, the arcs on them, AB, HΘ and LM, will also be similar.

So it is true, not only that the epicycle has traversed arc AB in the same time as the body has traversed arc EZ, but also that the body will have traversed arcs HΘ and LM on the eccentres in that same time; hence in every case it will be seen along the same line DMZΘ, according to the epicyclic [hypothesis] at point Z, according to the greater eccentre at point Θ, and according to the smaller eccentre at point M. The same will hold true in all positions.

A further consequence is that where the apparent distance of the body from apogee [at one moment] equals its apparent distance from perigee [at another], the equation of anomaly will be the same at both positions.

[Proof:] In the eccentric hypothesis [see Fig. 3.7], we draw the eccentric circle ABGD on centre E and diameter AEG through apogee A. We suppose the observer to be located at Z, and draw an arbitrary [chord] BZD through Z, and join EB and ED. Then the apparent positions [of the body at B and D] will be equal and opposite, that is the angle AZB from the apogee will be equal and opposite to angle GZD from the perigee; and the equation of anomaly will be the same [in both cases], since

$$BE = ED, \text{ and } \angle \ EBZ = \angle \ EDZ.$$

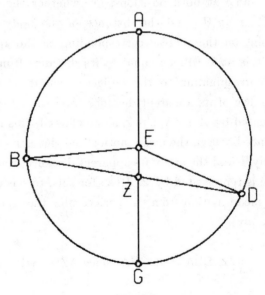

Fig. 3.7

So the arc [AB] of mean motion counted from the apogee A will exceed the arc of apparent motion (i.e. the arc subtended by angle AZB) by the same equation [equal to ∠ EBZ] as the arc of mean motion counted from the perigee G is exceeded by the arc of apparent motion (i.e. the [equal] arc subtended by ∠ GZD). For

∠ AEB > ∠ AZB, and ∠ GED < ∠ GZD.

In the epicyclic hypothesis [see Fig. 3.8] if, as before, we draw the concentre ABG on centre D and diameter ADC, and the epicycle EZH on centre A, draw an arbitrary line DHBZ, and join AZ and AH, then the arc AB representing the equation of anomaly will be the same at both positions, i.e. whether the body is at Z or at H. And the distance of the body from the point on the ecliptic corresponding to the apogee when it is at Z will be equal to its distance from the point corresponding to the perigee when it is at H. For the arc of its apparent distance from the apogee is represented by ∠ DZA, since, as we showed, this is the difference between the mean motion and the equation of anomaly.[57] And the arc of its apparent distance from the perigee is represented by ∠ ZHA (for this, too, is equal to the mean motion from the perigee plus the equation of anomaly).

But ∠ DZA = ∠ ZHA, since AZ = AH.

---

57   ∠ DZA = ∠ EAZ – ∠ ADZ.

Thus here too we conclude that the mean motion exceeds the apparent near the apogee (i.e. ∠ EAZ exceeds ∠ AZD) by the same equation (namely ∠ ADH) as the mean motion is exceeded by the (same) apparent motion (i.e. ∠ HAD by ∠ AHZ) near the perigee.

Q.E.D.

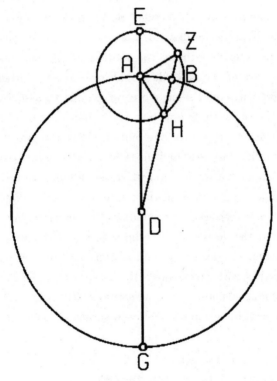

Fig. 3.8

4. On the apparent anomaly of the Sun[58]

Having set out the above preliminary theorems, we must add a further preliminary thesis concerning the apparent anomaly of the sun. This has to be a single anomaly, of such a kind that the time taken from least speed to mean shall always be greater than the time from mean speed to greatest, for we find that this accords with the phenomena. Now this could be represented by either of the hypotheses described above, though in case of the epicyclic hypothesis the motion of the sun on the apogee arc of the epicycle would have to be in advance. However, it would seem more reasonable to associate it with the eccentric hypothesis, since that is simpler and is performed by means of one motion instead of two.[59]

Our first task is to find the ratio of the eccentricity of the sun's circle, that is, the ratio which the distance between the centre of the eccentre and the centre of the ecliptic (located at the observer) bears to the radius of the eccentre. We must also find the degree of the ecliptic in which the apogee of the eccentre is located. These problems have been solved by Hipparchus with great care.[60] He assumes that the interval from spring equinox to summer solstice is 94½ days, and that the interval from summer solstice to autumnal equinox is 92½

---

58   See *HAMA* 57–8, Pedersen 144–9.

59   On the desirability of simplicity in hypotheses see III I.

60   Reading μετά πάσης σπουδῆς (with D, Ar) at H233, 1–2 for μετά σπονδῆς ('with care').

days, and then, with these observations as his sole data, shows that the line segment between the above-mentioned centres [of eccentre and ecliptic] is approximately 1/24th of the radius of the eccentre, and that the apogee is approximately 24½° (where the ecliptic is divided into 360°) in advance of the summer solstice. We too, for our own time, find approximately the same values for the times [taken by the sun to traverse] the above-mentioned quadrants, and for those ratios. Hence it is clear to us that the sun's eccentre always maintains the same position relative to the solsticial and equinoctial points.[61]

In order not to neglect this topic, but rather to display the theorem worked out according to our own numerical solution, we too shall solve the problem, for the eccentre, using the same observed data, namely, as already stated, that the interval from spring equinox to summer solstice comprises 94½ days, and that from summer solstice to autumnal equinox 92½ days. For our own very precise observations of the equinoxes and the summer solstice in the 463rd year from the death of Alexander confirm the day-totals in these intervals: as we said, (III 1), the autumnal equinox occurred on Athyr [III] 9, [139 Sept. 26], after sunrise, the spring

---

61   According to Ptolemy the sun's apogee (unlike those of the five planets, as it later turns out, IX 7) does not share in the motion of precession. The reproaches that have been cast on Ptolemy (e.g. by Manitius I 428–9) for failing to discover that the sun's apogee too has a motion through the ecliptic are unjustified. To do that he would have needed observations of the time of equinox and solstice far more accurate than those available (to the nearest ¼-day), and not only for his own time but also for an earlier time. See the papers by Rome[3] and Petersen and Schmidt for a mathematical demonstration of this.

equinox on Pachon [IX] 7 [140 March 22), after noon (thus the interval [between them] is 178¼ days), and the summer solstice on Mesore [XII] 11/12, [140 June 24/25], after midnight. Thus this interval, from spring equinox to summer solstice, comprises 94½ days, which leaves approximately 92½ days to complete the year; this number represents the interval from the summer solstice to the following autumnal equinox.[62]

[See Fig. 3.9.] Let the ecliptic be ABGD on centre E. In it draw two diameters, AG and BD, at right angles to each other, through the solsticial and equinoctial points. Let A represent the spring [equinox], B the summer [solstice], and so on in order.

Now it is clear that the centre of the eccentre will be located between lines EA and EB. For semi-circle ABG comprises more than half of the length of the year, and hence cuts off more than a semi-circle of the eccentre; and quadrant AB too comprises a longer time and cuts off a greater arc of the eccentre than quadrant BG. This being so, let point Z represent the centre of the eccentre, and draw the diameter through both centres and the apogee, EZH. With centre Z and arbitrary radius draw the sun's eccentre ΘKLM, and draw through Z lines

---

62   In III I the precise times of day given are '1 hour after sunrise', '1 hour after noon' and '2 hours after midnight'. Thus the precise intervals are 178¼ days and 94d 13h, leading to corrected figures of 94d 13h and 92d 11h for the intervals used in the computation. But see n.23 for the possibility that the time of solstice is '2 seasonal hours' (≈ 2/3 equinoctial hours). Even as small a change as 1 hour in an interval has an effect of about 1° in the location of the apogee (cf. Petersen and Schmidt 80–3 and Rome[3] 13–15).

NXO parallel to AG and PRS parallel to BD. Draw perpendicular ΘTY from Θ to NXO and perpendicular KFQ from K to PRS.

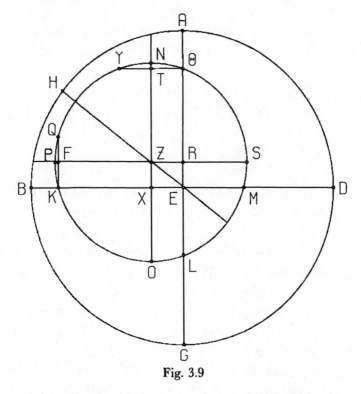

Fig. 3.9

Now since the sun traverses circle ΘKLM with uniform motion, it will traverse arc ΘK in 94½ days, and arc KL in 92½ days. In 94½ days its mean motion is approximately 93;9°, and in 92½ days 91;11°. Therefore

arc ΘKL = 184;20°

and, by subtraction of the semi-circle NPO [from arc ΘKL],

arc NΘ + arc LO [= 184;20° − 180°] = 4;20°

So arc ΘNY = 2 arc ΘN = 4;20° also,

∴ ΘY = Crd arc ΘNY ≈ 4;32^P ⎫ where the diameter of
and EX = ΘT = ½ΘY = 2;16^P ⎭ the eccentre = 120^P,

Now since arc ΘNPK = 93;9°,

and arc ΘN = 2;10° and quadrant NP = 90°, by subtraction, arc PK = 0;59°,

and arc KPQ = 2 arc PK = 1;58°.

∴ KFQ = Crd arc KPQ = 2;4^P, ⎫ where the diameter of
and ZX = KF = ½ KFQ = 1;2^P ⎭ the eccentre = 120^P,

And we have shown that EX = 2;16p in the same units.

Now since EZ2 = ZX2 + EX2,

EZ ≈ 2;29½^P where the radius of the eccentre = 60^P.

Therefore the radius of the eccentre is approximately 24 times the distance between the centres of the eccentre and the ecliptic.

Now, since EZ:ZX = 2;29½ : 1;2,

ZX will be about 49;46^P where hypotenuse EZ = 120^P. Therefore, in the circle about right-angled triangle EZX, arc ZX ≈ 49°.

∴ ∠ ZEX = { 49°° where 2 right angles = 360°°
24;30° where 4 right angles = 360°°.

82

So, since ∠ ZEX is an angle at the centre of the ecliptic, arc BH, which is the amount by which the apogee at H is in advance of the summer solstice at B, is also 24;30°.

Furthermore, since quadrants OS and SN are each 90°, and arc OL = arc ΘN = 2;10°,

and arc MS = 0;59°,

∴ arc LM = 86;51°,

and arc MΘ = 88;49°. But the sun in its uniform motion travels 86;51° in about 81/8 days, and 88;49° in about 901/8 days.

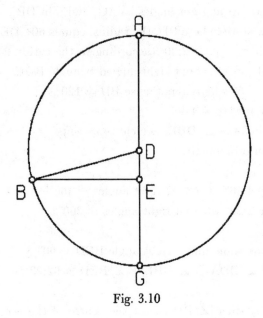

Fig. 3.10

Hence it is clear that the sun will traverse arc GD, which extends from the autumnal equinox to the winter solstice, in about 881/8 days, and arc DA, which extends from the winter solstice to the spring equinox, in about 901/8 days. The above conclusions are in agreement with what Hipparchus says.

Using these quantities, then, let us first see what the greatest difference between mean and anomalistic motions is, and at what points it will occur. [See Fig. 3.10.] Let the eccentric circle be ABG on centre D and diameter ADG through the apogee A, on which E represents the centre of the ecliptic.

Draw EB at right angles to AG, and join DB.

Now since, where BD, the radius, equals 60^p, DE, the eccentricity, equals 2;30^p (according to the ration 24:1),

in the circle about right-angled triangle BDE,

DE = 5^p where hypotenuse BD = 120^p,

and arc DE ≈ 4;46°.

Therefore ∠ DBE, which represents the greatest equation of anomaly,

$$= \begin{cases} 4;46°° \text{ where } 2 \text{ right angles} = 360°° \\ 2;23° \text{ where } 4 \text{ right angles} = 360°°. \end{cases}$$

In the same units, right angle BED = 90°,

and ∠ BDA = ∠ DBE + ∠ BED = 92;23°.

Thus, since ∠ BDA is at the centre of the eccentre and ∠ BED is at the centre of the ecliptic, we conclude that the greatest equation of anomaly is 2;23°, and the position where it occurs is 92;23° from the apogee, measured along the eccentre in uniform motion, and (as we proved earlier) a quadrant, or 90° [from the apogee], measured along the ecliptic in anomalistic motion. It is obvious from our previous results that in the opposite semi-circle[63] the mean speed and the greatest equation of anomaly will occur at 270° of apparent motion, and at 267;37° of mean motion on the eccentre.

We now want to use numerical computation, as we promised, to show that one derives the same quantities from the epicyclic hypothesis too, provided the same ratios are preserved in the way we explained.

[See Fig. 3.11] Let the circle concentric to the ecliptic be ABG on centre D and diameter ADG, and the epicycle circle EZH on centre A. From D draw a tangent to the epicycle, DZB, and join AZ. Then, as before, in the right-angled triangle ADZ, AD is 24 times AZ, so that, in the circle about right-angled triangle ADZ, AZ is, again, 5^P where hypotenuse AD is 120^P, and the arc on AZ is 4;46°.

$$\therefore \angle \text{ ADZ} = \left\{ \begin{array}{l} 4;46°° \text{ where 2 right angles} = 360°° \\ 2;23° \text{ where 4 right angles} = 360°. \end{array} \right.$$

Therefore the greatest equation of anomaly, namely arc AB, has been found to be 2;23° here too, in agreement

---

63   Reading ἡμικύκλιον (with D,Ar) for τμῆμα ('segment') at H239,12.

85

with [the previous result], and the arc of anomalistic motion is 90°, since it is represented by the right angle AZD, while the arc of mean motion, which is represented by ∠ EAZ, is again 92;23°.

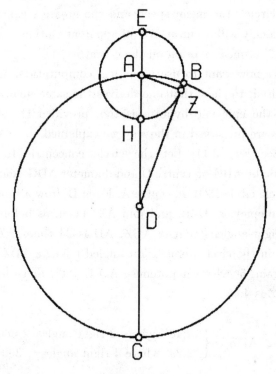

Fig. 3.11

Book IV

1. The kind of observations which one must use to examine Lunar phenomena[64]

In the preceding book we treated all the phenomena associated with the sun's motion. We now begin our discussion of the moon, as is appropriate to the logical order. In doing so we think it our first duty not to take a naive or arbitrary approach in our use of the relevant observations. Rather, to establish our general notions [on this topic], we should rely especially on those demonstrations which depend on observations which not only cover a long period, but are actually made at lunar eclipses. For these are the only observations which allow

---

64   On Chs 1–3 see *HAMA* 68–73, 308–15, Pedersen 160–4.

one to determine the lunar position precisely: all others, whether they are taken from passages [of the moon] near fixed stars, or from [sightings with] instruments, or from solar eclipses, can contain a considerable error due to lunar parallax. It is only for particular further developments [of the theory] that we should use these other kinds of observations for our investigations. For the distance between the sphere of the moon and the centre of the earth, unlike the distance to the ecliptic, is not so great that the earth's bulk has the ratio of a point to it. Hence it necessarily follows that the straight line drawn from the centre of the earth (which is the centre of the ecliptic) through the centre of the moon[65] to a point on the ecliptic, which determines the true position ([as it does] for all bodies), does not in this case always coincide, even sensibly, with the line drawn from some point on the earth's surface, that is, the observer's point of view, to the moon's centre, which determines its apparent position. Only when the moon is in the observer's zenith do the lines from the earth's centre and the observer's eye through the moon's centre to the ecliptic coincide. But when the moon is displaced from the zenith position in any way whatever, the directions of

---

65   Reading ἀπό τοῦ κέντρον τῆς γῆς τουτεστι τοῦ ξῳδιακού διά τοῦ κέντρου τῆς σελήνης (with D, Ar) for ἀπό τοῦ κέντρου τῆς σελήνης ('the straight line drawn from the moon's centre', which is nonsense) at H266,5. The error in most Greek mss. is due to haplography, and is an important indication that all except D and its descendants come from a single (?Byzantine) ms. Corrected by Manitius.

the above lines become different, and hence the apparent position cannot be the same as the true, but [differs from it], as the [line through] the observer's eye assumes various positions with respect to the line drawn through the centre of the earth, [by an amount] proportional to the varying angle of inclination [between the two lines].

This is the reason why in the case of solar eclipses, which are caused by the moon passing below and blocking [the sun] (for when the moon falls into the cone from the observer's eye to the sun it produces the obscuration which lasts until it has passed out [of the cone] again), the same[66] eclipse does not appear identical, either in size or in duration,[67] in all places. For the moon does not produce obscuration for all observers, for the reasons stated above, and [even for those for whom it does produce obscuration] does not appear to obscure the same parts of the sun [for all alike]. Whereas in the case of lunar eclipses there is no such variation due to parallax, since the observer's position is not a contributory cause to what happens at a lunar eclipse. For the moon's light is at all times caused by the illumination from the sun. Thus when it is diametrically

---

66   Reading τάς αὐτάς (with D, Ar) for tautac; ('these eclipses') at H267,4.
Corrected by Manitius.

67   'duration': the Greek has the vague 'times' (τοίς χρόνοίς). This is elucidated by H268,l
τοίς τῶν διαστασεων χρόνοίς 'the duration of the intervals [of partial and total phases]."'
Ptolemy may also be alluding, in both places, to the fact that the actual moments of e.g.
the beginning or middle of a solar eclipse are different at different places, and by an amount
which does not correspond directly to the difference in longitude.

opposite to the sun, it normally appears to us as lighted over its whole surface, since the whole of its illuminated hemisphere is turned towards us as well [as towards the sun] at that time. However, when its position at opposition is such that it is immersed in the earth's shadow-cone (which revolves with the same speed as the sun, but opposite it), then the moon loses the light over a part of its surface corresponding to the amount of its immersion, as the earth obstructs the illumination by the sun. Hence it appears to be eclipsed for all parts of the earth alike, both in the size [of the eclipse] and the length of the intervals [of the various phases].

Now to establish our general theory we need to use true, and not apparent, positions of the moon; for the ordered and regular must necessarily precede and serve as a foundation for the disordered and irregular. So, for the above reasons, we declare that we must not use, for this purpose, observations of the moon into which the observer's position enters, but only lunar eclipse observations, since [only] in these does the observer's position have no effect on the determination of the moon's position. For it is obvious that, if we find the point on the ecliptic which the sun occupies at the time of mid-eclipse (which is, as accurately as we can determine, the moment at which the moon's centre is diametrically opposite the sun's in longitude), then at the same time of mid-eclipse the precise position of the moon's centre will be the point diametrically opposite.

THE BOOK OF ASTRONOMY IN ANTIQUITY

2. On the periods of the moon

The above may serve as an outline of the kind of observations which must be examined to determine the general theory of the moon. We shall now endeavour to describe the method which was used by the ancients in their attempts at establishing a [lunar] theory, and which we will find a most convenient tool in deciding which hypotheses accord with the phenomena. The moon's motion appears anomalistic both in longitude and in latitude: the time it takes to traverse the ecliptic is not constant, and neither is the time it takes to return to the same latitude.[68] Now unless one finds the period of its return in anomaly it is, necessarily, impossible to determine the period of the other motions [in longitude and latitude]. However, from individual observations it is apparent that the moon's mean speed can occur in any part of the ecliptic, as can its greatest speed and its least speed, and that it can reach its greatest northern or southern latitude, or appear exactly in the ecliptic, anywhere, too. Hence the ancient astronomers, with good reason, tried to find some period in which the moon's motion in longitude would always be the same, on the grounds that only such a period could produce a return in anomaly. So they compared observations of lunar eclipses (for the reasons mentioned above), and tried to

---

68   Reading κατά πλάτος (with D) for κατά τό πλάτος at H269,9.

see whether there was an interval, consisting of an integer number of months, such that, between whatever points one took that interval of months,[69] the length in time was always the same, and so was the motion [of the moon] in longitude, [i.e.] either the same number of integer revolutions, or the same number of revolutions plus the same arc. The even more ancient [astronomers] used the somewhat crude estimate that such a period could be found in 6585⅓ days. For they saw that in that interval occurred approximately 223 lunations, 239 returns in anomaly, 242 returns in latitude, and 241 revolutions in longitude plus 10⅔°, which is the amount the sun travels beyond the 18 revolutions which it performs in the above time (that is when the motion of sun and moon is measured with respect to the fixed stars). They called this interval the 'Periodic', since it is the smallest single period which contains (approximately) an integer number of returns of the various motions.[70] In order to obtain a period with an integer number of days, they tripled the 6585⅓ days, obtaining 19756 days, which they called 'Exeligmos'. Similarly, by tripling the other numbers, they obtained 669 lunations, 717 returns in anomaly, 726 returns in latitude, and 723 revolutions in

---

69   'months' here means 'true synodic months'. This is generally true throughout the Almagest (except where the context makes it obvious that the reference is strictly calendaric). In the translation I usually make the meaning explicit.

70   This period, generally, but wrongly, called 'Saros' in modern times (see Neugebauer[I]), was well-known in Babylonian astronomy. See *HAMA* 497 ff. We do not know to whom Ptolemy refers by 'the even more ancient people', except that they are earlier than Hipparchus.

longitude plus 32°, which is the amount the sun travels beyond its 54 revolutions.[71]

However, Hipparchus already proved, by calculations from observations made by the Chaldaeans and in his time, that the above relationships were not accurate. For from the observations he set out he shows that the smallest constant interval defining an ecliptic period in which the number of months and the amount of [lunar] motion is always the same, is 126007 days plus 1 equinoctial hour. In this interval he finds comprised 4267 months, 4573 complete returns in anomaly, and 4612 revolutions on the ecliptic less about 7½°, which is the amount by which the sun's motion falls short of 345 revolutions (here too the revolution of sun and moon is taken with respect to the fixed stars). (Hence, dividing the above number of days by the 4267 months, he finds the mean length of the [synodic] month as approximately 29, 31, 50, 8, 20 days). He shows, then, that the corresponding interval between two lunar eclipses is always precisely the same when they are taken over the above period [l26007dlh]. So it is obvious that it is a period of return in anomaly, since [from whatever eclipse it begins], it always contains the same number [4267] of months, and 4611 revolutions in longitude plus 352½°, as determined by its syzygies with the sun.

---

71   The ἐξελιγμός (meaning 'turn of the wheel') is also mentioned by Geminus (Cap. XVIII. ed. Manitius pp. 200–2), who gives exactly the same numbers as Ptolemy, including the excess, in sidereal longitude of 32°.

But if one were to look for the number of months [which always cover the same time-interval], not between two lunar eclipses, but merely between one conjunction or opposition and another syzygy of the same type, he would find an even smaller integer number of months containing a return in anomaly, by dividing the above numbers by 17 (which is their only common factor). This produces 251 months and 269 returns in anomaly.

However, it was found that the above period [of 126007d1h] did not contain an integer number of returns in latitude too. For it was apparent that the [pairs of] corresponding eclipses exhibited equality only with respect to the interval [between the pair] in time and revolution in longitude, but not with respect to the size and type of the obscuration,[72] which is the criterion for [a return in] latitude. Nevertheless, having already determined the period of return in anomaly, Hipparchus again adduces intervals containing [an integer number of] months which have at each end eclipses which were identical in every respect, both in size and in duration [of the various phases], and in which there was no difference due to the anomaly. Thus it is apparent that there is a return in latitude too. He shows that such a period is contained in 5458 months and 5923 returns in latitude.[73]

---

72 By 'type' Ptolemy means whether the obscuration begins from the north or south of the lunar disk.

73 Ptolemy's account here is not historically accurate. In fact Hipparchus took from Babylonian sources the parameters (1) I synodic month = 29; 31, 50, 8, 20°, (2) 251 synodic months = 269 anomalistic months, and (3) 5458 synodic months = 5923 returns in latitude (Kugler, Babylonische Mundrechnung 4–46). Multiplying [2] by 17, he constructed an

That, then, is the method which our predecessors used for the determination of such [periods]. It is not simple or easy to carry out, but demands a great deal of extraordinary care, as we can see from the following considerations.[74] Let us grant that [two] intervals [between pairs of eclipses] are found to be precisely equal in time. In the first place, this is no use to us unless the sun too exhibits no effect due to anomaly, or exhibits the same over both intervals: for if this is not the case, but instead, as I said, the equation of anomaly has some effect, the sun will not have travelled equal distances over [the two] equal time-intervals, nor, obviously, will the moon. For example, let us suppose that each of the two intervals being compared comprises half a year beyond the same number of complete years, and that in this time the motion of the sun in the first interval starts from the position of mean speed in Pisces, and in the second interval from the position of mean speed in Virgo.[75] Then over the first interval the sun will have traversed about 4¾° less than a semi-circle [beyond complete revolutions], but over the second

---

eclipse-period (Aaboe [1955], whence *HAMA* 310–2). An input of some value for the length of the year produced the solar motion over this period. rounded by Hipparchus to the nearest ¼ -sign (on which see Neugebauer[2], 251). Then Hipparchus confirmed (not derived, as Ptolemy says) the above by comparison of eclipses from his own time with Babylonian ones 345 years earlier (see Toomer[11] for the method and identification of the eclipses he used).

74   The following is well explained and illustrated by Neugebauer, *HAMA* 71–2.

75   That is, from the positions where the equation of anomaly reaches its positive maximum (Pisces) and negative maximum (Virgo). Illustrated by *HAMA* Fig. 59 p. 1223.

about 4¾° more than a semi-circle. Thus the moon too will have traversed over the first interval 175¼° beyond complete revolutions and over the second 184⅙°, although both intervals cover an equal time. Therefore we define as the first necessary condition [for a return in lunar anomaly] that the intervals must exhibit one of the following characteristics with respect to the sun:

1. It must complete an integer number of revolutions [in both intervals]; or
2. traverse the semi-circle beginning at the apogee over one interval and the semi-circle beginning at the perigee over the other; or
3. begin from the same point [of the ecliptic] in each interval; or
4. be the same distance from apogee (or perigee) at the first eclipse of one interval as it is at the second eclipse of the other interval, [but] on the other side.[76]

For only under one of these conditions will there be no effect due to the anomaly, or the same effect over both intervals, so that the arc traversed beyond complete revolutions over one interval is equal to that traversed over the other, or even equal to the mean motion of the sun [over the intervals] as well.

76   That is, if the sun has an anomaly of α° at the beginning of the first interval, it must have an anomaly of (360 - α)° at the end of the second interval. This situation (and the others listed here) is illustrated by *HAMA* Fig. 60 p. 1223.

Secondly, it is our opinion that we must pay no less attention to the moon's [varying] speed.[77] For if this is not taken into account, it will be possible for the moon, in many situations, to cover equal arcs in longitude in equal times which do not at all represent a return in lunar anomaly as well. This will come to pass

1. if in both intervals the moon starts from the same speed (either both increasing or both decreasing), but does not return to that speed; or
2. if in one interval it starts from its greatest speed and ends at its least speed, while in the other interval it starts from its least speed and ends at its greatest speed; or
3. if the distance of [the position of] its speed at the beginning of one interval is the same distance from the [position of] greatest or least speed as [the position of] its speed at the end of the other interval, [but] on the other side.[78]

---

77   δρόμος is often used in early Greek astronomy for the (varying) amount which the moon travels in one day. The earliest example seems to be the 'Eudoxus' papyrus (ed. Blass p. 14). Where Ptolemy uses δρόμος for the moon (e.g. V 2, H355,14; V 3, H361,16) 'speed' seems the best translation.

78   Illustrated (in the order [1], [3], [2]) by *HAMA* Fig. 61 p. 1224, which utilizes the lunar epicycle model. One must presume that Ptolemy avoids talking in geometrical terms (which is the most convenient way to visualize the situation) because he has not yet established a lunar model. However, it is hard to give any sense to ἐκατέρωθεν (literally 'on opposite sides'; translated here as 'on the other side') which does not involve an epicycle model.

In each of these situations there will again be either no effect or the same effect [in both intervals] of the lunar anomaly, and hence equal increments in longitude will be produced [over both intervals], but there will be no return in anomaly at all. So the intervals adduced must avoid all the above situations if they are to provide us directly with a period of return in anomaly. On the contrary, we should select intervals [the ends of which are situated] so as to best indicate [whether the interval is or is not a period of anomaly], by displaying the discrepancy [between two intervals] when they do not contain an integer number of returns in anomaly. Such is the case when the intervals begin from speeds which are not merely different, but greatly different either in size or in effect. By 'in size' I mean when in one interval [the moon] starts from its least speed and does not end at the greatest speed, while in the other it starts from its greatest speed and does not end at its least speed. For in this case, unless the intervals contain an integer number of revolutions in anomaly, the difference in the increments in longitude over the two intervals will be very great; when the increment in anomaly is about one or three quadrants of a revolution, the intervals will differ by twice the [maximum] equation of anomaly. By 'in effect' I mean when [the moon] starts from mean speed in both positions, not, however, from the same mean speed, but from the mean speed during the period of increasing speed at one interval, and from that during the period of decreasing speed at the other. Here too, if there is not a return in anomaly, there will be a great difference in the increment in longitude [over the two intervals]; again, when the increment in anomaly is one or three quadrants of a revolution, the difference will again amount to twice the [maximum] equation of anomaly, and when the increment in anomaly is a semi-circle, the difference will be four times that amount.[79]

That is why, as we can see, Hipparchus too used his customary extreme care in the selection of the intervals adduced for his investigation of this question: he used [two intervals], in one of which the moon started from its greatest speed and did not end at its least speed, and in the other of which it started from its least speed and did not end at its greatest speed. Furthermore he also made a correction, albeit a small one, for the sun's equation of anomaly, since the sun fell short of an integer number of revolutions by about ¼ of a sign, and this sign was different, and produced a different equation of anomaly, in each of the two intervals.[80]

We have made the above remarks, not to disparage the preceding method of determining the periodic returns, but to show that, while it can achieve its goal

---

79   These two situations (of maximum effect due to the anomaly when there is not a return in anomaly) are illustrated by *HAMA* Fig. 62 p. 1225.

80   On the eclipses used by Hipparchus see Toomer[11].

if applied with due care and the appropriate kind of calculations, if any of the conditions we set out above are omitted from consideration, even the least of them, it can fail utterly in its intended effect; and that, if one does use the proper criteria in making one's selection of observational material, it is difficult to find corresponding [pairs of eclipse] observations which precisely fulfil all the required conditions.

In any case, when we take the above periodic returns, as determined by Hipparchus' calculations, we find that the period [containing an integer number] of months has, as we said, been calculated as correctly as possible, and has no perceptible difference from the true value. But there is an error in the periods of anomaly and latitude, so considerable as to become quite apparent to us from the procedures we devised to check these values in simpler and more practical ways; we shall soon explain these, in connection with our demonstration of the size of the lunar anomaly. But first, for convenience [of calculation] in what follows, we set out the individual mean motions [of the moon] in longitude, anomaly and latitude, in accordance with the above periods of their returns, and [also the mean motions] calculated on the basis of the corrections which we shall derive later.[81]

---

81   Ptolemy's corrections to the mean motions in anomaly and latitude, given below, are justified at IV 7 and IV 9.

5. That in the simple hypothesis of the moon, too, the same phenomena are produced by both eccentric and epicyclic hypotheses[82]

Our next task is to demonstrate the type and size of the moon's anomaly. For the time being we shall treat this as if it were single and invariant.[83] It is apparent that this anomaly, namely the one with a period corresponding to the above period of return, is the only one which our predecessors (just about all of them) have hit upon. Later, however, we shall show that the moon also has a second anomaly, linked to its distance from the sun; this [second anomaly] reaches a maximum round about both [waxing and waning] half-moons, and goes through its period of return twice a month, [being zero] precisely at conjunction and opposition.[84] We adopt this order of procedure in our demonstration because it is impossible to determine the second [anomaly] apart from the first, which is always combined with it, whereas the first can be found apart from the second, since it is determined from lunar eclipses, at which there is no perceptible effect of the anomaly connected with [the distance

---

82 See Pedersen 166–7.

83 Reading καί τῆς αὐτῆς (with BD) for ταύτης ('as if this were single') at H294,6. Ar read ταύτης.

84 Reference to V 2–4.

from] the sun. In this first part of our demonstrations we shall use the methods of establishing the theorem which Hipparchus, as we see, used before us.[85] We too, using three lunar eclipses, shall derive the maximum difference from mean motion and the epoch of the [moon's position] at the apogee, on the assumption that only this [first] anomaly is taken into account, and that it is produced by the epicyclic hypothesis. It is true that the same phenomena would result from the eccentric hypothesis, but we shall find the latter more suitable to represent the second anomaly, which is connected with the sun, when we come to combine both anomalies. However, the same phenomena will in all cases result from both the hypotheses we have described, whether, as in the situation described for the sun, the period of return in anomaly and the period of return in the ecliptic [i.e. in longitude] are both equal, or whether, as in the case of the moon, they are unequal, provided only that the ratios [of epicycle to deferent and eccentricity to eccentre] are taken as identical. We can see this from the following, in which we use the above-mentioned simple anomaly of the moon for our examination.

Since the moon completes its return with respect to the ecliptic sooner than its return with respect to this anomaly, it is clear that, in the epicyclic hypothesis, over

85   On Hipparchus' determination of the lunar parameters see further IV 11, Toomer[8] and Toomer[2].

a given period of time, the epicycle will always traverse a greater arc[86] of the circle concentric to the ecliptic than the arc of the epicycle traversed by the moon in the same time; in the eccentric hypothesis, the arc traversed by the moon on the eccentre will be similar to the arc traversed by it on the epicycle [in the epicyclic hypothesis], while the eccentre will move about the centre of the ecliptic in the same direction as the moon by an amount equal to the increment of the motion in longitude over the motion in anomaly [in the same time] (this corresponds to the increment of the arc of the deferent over the arc of the epicycle [in the epicyclic hypothesis]). In this way we can preserve the equality of the periods of both motions [i.e. in longitude and anomaly], as well as equality of the ratios, in both hypotheses.

With the above as a necessary basis (as is obvious from logic), let [Fig. 4.1] the circle concentric with the ecliptic be ABG on centre D and diameter AD, and let the epicycle be EZ on centre G. Let us suppose that when the epicycle was at A, the moon was at E, the apogee of the epicycle, and that in the same time as the epicycle has traversed arc AG, the moon has traversed arc EZ. Join ED, GZ.

Then, since arc AG > arc EZ,

cut off arc BG ∥ arc EZ, and join BD.

Then it is clear that, in the same time, the eccentre

---

86   'a greater arc': literally 'an arc greater than the one similar to [the arc].

103

will have moved through ∠ ADB, which represents the difference between the two motions, and its centre and apogee will tie along line BD.

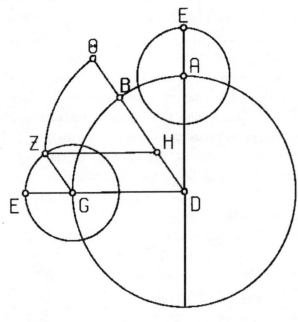

Fig. 4.1

This being so, let DH = GZ. Join ZH, and with centre H and radius HZ draw the eccentre ZΘ. I say, that

ZH:HD = DG:GZ,

and that in this hypothesis too the moon
will be at point Z, i.e.

arc ZEΘ ‖ arc EZ.

104

[Proof:] Since ∠ BDG = ∠ EGZ, GZ is parallel to DH.
But GZ = DH [by construction].
Therefore ZH too is equal and parallel to GD.[87]
∴ ZH:HD = DG:GZ.
Furthermore, since DG is parallel to HZ,
∠ GDB = ∠ ZHΘ
and, by hypothesis, ∠ GDB = ∠ EGZ.
∴ arc ZΘ ∥ arc EZ.

Therefore the moon has reached point Z in the same time according to either hypothesis, since the moon itself has traversed arc EZ on the epicycle and arc ΘZ on the eccentre, which we have shown to be similar, while the epicycle centre has moved through arc AG, and the centre of the eccentre through arc AB, which is the increment of arc AG over arc EZ.

Q.E.D.

Moreover, even if [the members of] the ratios are unequal, and the eccentre is not the same size as the deferent, the same phenomena will result, provided the ratios are similar, as will be clear from the following.

Draw each of the hypotheses in a separate figure. Let [Fig. 4.2] the circle concentric to the ecliptic be ABG on centre D and diameter AD, and the epicycle EZ on centre G. Let the moon be at Z. Let [Fig. 4.3] the eccentre be HΘK on centre L and diameter ΘLM, with the centre of the ecliptic at M. Let the moon be at K. In the first figure join

---

87   Euclid I 33: straight lines joining equal and parallel lines are themselves equal and parallel.

DGE, GZ, DZ, and in the second figure join HM, KM, KL. Let DG:GE = ΘL:LM.

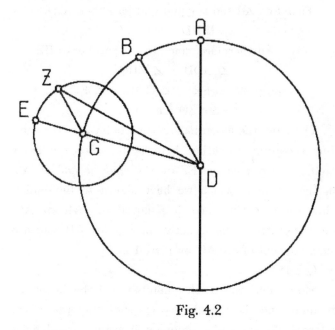

Fig. 4.2

Let us suppose that in the same time as the epicycle has moved through ∠ ADC, the moon has again moved through ∠ EGZ, the eccentre through ∠ HMΘ, and the moon, again, through ∠ ΘLK.

Therefore, because of the assumed relationship between the motions,

∠ EGZ = ∠ ΘLK,

and ∠ ADG = ∠ HMΘ + ∠ ΘLK.

106

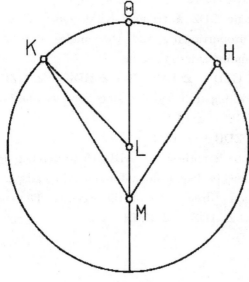

Fig. 4.3

This being so, I say that the moon will again appear to have traversed an equal arc in the same time according to either hypothesis, i.e.

∠ ADZ = ∠ HMK

(for at the beginning of the time-interval the moon was at the apogee and appeared along lines DA and MH, while at the end it was at points Z and K and appeared along lines ZD and MK).

[Proof:] Let arc BG again be similar to arc ΘK (or arc EZ). Join BD.

Then, since DG:GZ = KL:LM,

and the angles at G and L are equal,

triangle GDZ ⫴ triangle KLM (sides about equal angles proportional), and the angles opposite the corresponding sides are equal.

∴ ∠ GZD = ∠ LMK. But ∠ BDZ = ∠ GZD,

for GZ is parallel to BD, since, by hypothesis,

∠ ZGE = ∠ BDG.

∴ ∠ ZDB = ∠ LMK.

But, by hypothesis, ∠ ADB, the difference between the motions [in longitude and anomaly] equals ∠ HMΘ, the motion of [the centre of] the eccentre. Therefore, by addition, ∠ ADZ = ∠ KMH.

Q.E.D.

Book V

1. On the construction of an 'astrolabe' instrument[88]

As far as concerns the [moon's] syzygies with the sun at conjunction and opposition, and the eclipses which occur at such syzygies, we find that the hypothesis set out above for the first, simple anomaly is sufficient, even if we employ it just as it is, without any change. But for particular positions [of the moon] at other sun-moon configurations one

---

88   On the instrument described in this chapter the only good discussion is that of Rome[4], to which the reader is referred for all details of its construction and use. My Fig. F is based on the drawing there. The numbers and letters designating the rings and other parts of the instrument also follow Rome's notation. In modern terms, it is an 'armillary sphere'. The adjective 'astrolabe' applied to it and to its parts simply means 'for taking the [the position of] the stars', and has nothing to do with the instrument to which the name 'astrolabe' is now usually applied (on which see *HAMA* II 868–79). The latter was called the 'small astrolabe' by Théon of Alexandria: see Rome[l] I 4 n.0; by Ptolemy it was apparently called 'horoscopic instrument' (see *HAMA* II 866).

will find that it is no longer adequate, since as we said, we have discovered that there is a second lunar anomaly, related to its distance from the sun. This anomaly is reduced to the first [i.e. becomes zero] at both syzygies, and reaches a maximum at both quadratures. We were led to awareness of and belief in this [second anomaly] by the observations of lunar positions recorded by Hipparchus,[89] and also by our own observations, which were made by means of an instrument which we constructed for this purpose. The makeup of the instrument is as follows.

We took two rings of an appropriate size, with their surfaces precisely turned on the lathe so as to be squared off [i.e. with rectangular cross-sections], equal and similar to each other in all dimensions.
We joined them together at diametrically opposite points, so that they were fixed at right angles to each other, and their corresponding surfaces coincided: thus one of them [Fig. F,3] represented the ecliptic, and the other [Fig. F,4] the meridian through

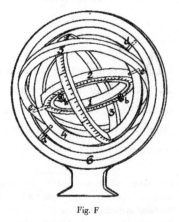

Fig. F

---

89 Examples of these are preserved at V 3 and V 5. It is notable that these are the latest three known observations of Hipparchus. The obvious conclusion is that towards the end of his career he suspected that the 'simple' lunar hypothesis was inadequate for positions outside the syzygies, and was making observations to check this.

the poles of the ecliptic and the equator [i.e. a colure]. On the latter, using the side of the [inscribed] square [as measure], we marked the points representing the poles of the ecliptic, and pierced each point with a cylindrical peg [Fig. F,e,e] projecting beyond both outer and inner surfaces. On the outer [projections] we pivoted another ring [Fig. F,5] the concave [inner] surface of which fitted closely on the convex [outer] surface of the two joined rings, in such a way that it could move freely about the above-mentioned poles of the ecliptic in the longitudinal direction. Similarly we pivoted another ring [Fig. F,2] on the inner [projections]; this too fitted the two [joined] rings closely, its convex surface to their concave, and, like the outer ring, moved freely in longitude about the same poles. We marked on this inner ring, and also on the ring representing the ecliptic, the divisions indicating the standard 360 degrees of the circumference, and as small subdivisions of a degree as was practical. Then we fitted snugly inside the inner of the two [movable] rings another thin ring [Fig. F,1] with sighting-holes [Fig. F,b,b] projecting from it at diametrically opposite points. [This ring was constructed] so that it could move laterally in the plane of the ring it was fitted into, towards either of the above-mentioned poles, in order to allow observation of the variation in latitude.

Having completed the above construction, we marked off from both poles of the ecliptic, on the ring representing the circle through both poles [Fig. F,4], an arc equal to the distance between the poles of ecliptic and equator (as

111

determined above). At the ends of these arcs (which were, again, diametrically opposite) we again inserted pivots [Fig. F;d,d], attaching them to a meridian ring [Fig. F,6] similar to that[90] described at the beginning of this treatise for making observations of the arc of the meridian between the solsticial points. This meridian ring was set up in the same position as the earlier one, perpendicular to the plane of the horizon and at an elevation of the pole appropriate for the place in question, and also parallel to the plane of the actual meridian [at that place]. Thus the inner rings [Fig. F,4 etc.] were set up so as to revolve about the poles of the equator, from east to west, following the first motion of the universe.

Once we had set up the instrument in the way described, whenever we had a situation in which both sun and moon could be observed above the earth at the same time, we set the outer astrolabe ring [Fig. F,5] to the graduation [on the ecliptic ring, fig. F,3] marking, as nearly as possible, the position of the sun at that moment. Then we rotated the ring through the poles [Fig. F,4] until the intersection [of outer astrolabe ring and ecliptic ring] marking the sun's position was exactly facing the sun, and thus both the ecliptic ring [Fig. F,3] and the [ring] which goes through the poles of the ecliptic [Fig. F,5] cast its shadow exactly on itself:[91] Or, if we

---

90   Reading τώ εν ἀρχή τῆς συντάξεως αποδεδειγμένω (with D,Ar) for τῶν ἐν ἀρχῆ τῆς συντάξεως ὑποδεδειγμένων (which is untranslatable) at H353,1–2.

91   According to Ptolemy's instructions, one has to compute the solar longitude, set the outer astrolabe ring (Fig. F, 5) to that position on the ecliptic ring (Fig. F, 3), and then, keeping the two in that position relative to each other, swing both until one can sight the sun along the outer astrolabe ring. Both rings should then shade themselves. Theoretically,

were using a star as sighting [i.e. orienting] object, we set the outer [astrolabe] ring to the position assumed for that star on the ecliptic-ring, [and then rotated the ring Fig. F,4 to such a position] that when we applied one eye to one face of the outer ring [Fig. F,5] the star appeared fastened, so to speak, to both [nearer and farther] surfaces of that face,[92] and thus was sighted in the plane through them. Then we rotated the other, inner astrolabe ring [Fig. F,2] towards the moon (or any other object we desired) so that the moon (or any other desired object) was sighted through both sighting-holes on the inmost ring at the same time as the sun (or the other sighting-star) was being sighted [as described above].

In this way we read off the position [of the moon or any other desired object] in longitude on the ecliptic, from the graduation occupied by the inner [astrolabe] ring [Fig. F,2] on the ring representing the ecliptic [Fig. F,3], and its

---

even without knowing the sun's position, one could set up the instrument by sighting the sun along the outer astrolabe ring and then moving the ecliptic ring relative to the latter until it shaded itself.

92   Reading ὥσπερ κεκολλημένος ἀμφοτέραις αὐτῆς ταῖς ἐπιφανείαις for καί διά τῆς ἀπεναντίον καί παραλλήλου τού κύκλου πλευράς ωσπερ κεκολλ ημένος ἀμφοτέραις αὐτῶν ταῖς ἐπιφανείαις at H353,24–354,l. The latter would mean 'when we applied one eye to the [nearer] face of the outer ring and [looked] along the opposite, parallel face of the ring, the star appeared fastened, so to speak, to the surfaces of both those faces'. The words καί δια ... πλευρας are a foolish explanatory interpolation by someone who misinterpreted ἀμφοτέραις to mean 'the opposite faces' of the ring instead of the two parts of the same face nearer to and farther from the eye'; then αὐτῆς (referring to τῆέτέρα τῶν πλευρῶν) was changed to αὐτῶν (referring to both πλευραί), or possibly αὐτῶν was simply interpolated. Quite apart from the technical problem, the text as printed by Heiberg is extraordinarily clumsy. The interpolation is quite early, since it is also in the Arabic tradition. Pappus' commentary to the passage betrays no hint that he read the interpolation, but is not sufficiently close to the Almagest to allow us to say that he did not.

deviation to north or south [of the ecliptic] along the circle through the poles of the ecliptic, from the graduations of the inner astrolabe ring [Fig. F,2]; the latter is given by the distance between the mid-point of the upper[93] sighting-hole on the inmost rotating ring [Fig. F,l] and the line drawn through the centre of the ecliptic ring.

When this type of observation was made without further analysis, it was found, both from the observations recorded by Hipparchus and from our own, that the distance of the moon from the sun was sometimes in agreement with that calculated from the above [simple] hypothesis, and sometimes in disagreement, the discrepancy being at some times small and at other times great. But when we paid more attention to the circumstances of the anomaly in question, and examined it more carefully over a continuous period, we discovered that at conjunction and opposition the discrepancy [between observation and calculation] is either imperceptible or small, the difference being of a size explicable by lunar parallax; at both quadratures, however, while the discrepancy is very small or nothing when the moon is at apogee or perigee of the epicycle, it reaches a maximum when the moon is near its mean speed and [thus] the equation of the first anomaly is also a maximum; furthermore, at either quadrature, when the first anomaly is subtractive the moon's observed position is at an even smaller longitude than that calculated by

---

93   'upper': literally 'above the earth'. Since the centre of all the rings represents the centre of the earth, the sight nearer the observer's eye is notionally 'below the earth', the other 'above the earth'.

subtracting the equation of the first anomaly, but when the first anomaly is additive its true position is even greater [than that calculated by adding the equation of the first anomaly], and the size of this discrepancy is closely related to the size of the equation of the first anomaly. From these circumstances alone we could see that we must suppose the moon's epicycle to be carried on an eccentric circle, being farthest from the earth at conjunction and opposition, and nearest to the earth at both quadratures. This will come about if we modify the first hypothesis along somewhat the following lines.

Imagine the circle (in the inclined plane of the moon) concentric with the ecliptic moving in advance, as before, (to represent the [motion in] latitude) about the poles of the ecliptic with a speed equal to the increment of the motion in latitude over the motion in longitude. Imagine, again, the moon traversing the so-called epicycle (moving in advance on its apogee arc) with a speed corresponding to the return of the first anomaly. Now, in this inclined plane, we suppose two motions to take place, in opposite directions, both uniform with respect to the centre of the eliptic: one of these carries the centre of the epicycle towards the rear through the signs with the speed of the motion in latitude, while the other carries the centre and apogee of the eccentre, which we assume located in the same [inclined] plane, (the centre of the epicycle will at all times be located on this eccentre), in advance through [i.e. in the reverse order of] the signs) by an amount corresponding to the difference between the motion in latitude and the double elongation (the elongation being

115

the amount by which the moon's mean motion in longitude exceeds the sun's mean motion). Thus, to give an example, in one day the centre of the epicycle traverses about 13;14° in motion of latitude towards the rear through the signs, but appears to have traversed 13;11° in longitude on the ecliptic, since the whole inclined circle [of the moon] traverses the difference of 0;3° in the opposite direction, [i.e.] in advance; [meanwhile] the apogee of the eccentre, in turn, travels 11;9° in the opposite direction, (again in advance): this is the amount by which the double elongation, 24;23°, exceeds the motion in latitude, 13;14°. The combination of both of these motions, which take place in opposite directions, as we said, about the centre of the ecliptic, will produce the result that the radius carrying the centre of the epicycle and the radius carrying the centre of the eccentre will be separated by an arc which is the sum of 13;14° and 11;9°, and is twice the amount of the elongation (which is approximately 12;11½°). Hence the epicycle will traverse the eccentre twice during a mean [synodic] month. We assume that it returns to the apogee of the eccentre at mean conjunction and opposition.

In order to illustrate the details of the hypothesis, imagine [Fig. 5.1] the circle in the moon's inclined plane concentric with the ecliptic as ABGD on centre E and diameter AEG. Let the apogee of the eccentre, the centre of the epicycle, the northern limit, the beginning of Aries and the mean sun [all] be located at point A at the same moment. Then I say that in the course of one day the whole [inclined] plane moves in advance from A towards D about centre E, by about 3': thus the northern limit (which is [still represented by] A) reaches ⊬ 29;57°. The two opposite motions are carried

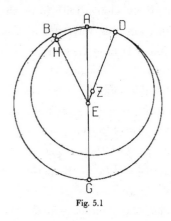

Fig. 5.1

out by the radius corresponding to EA [moving] uniformly about E, the centre of the ecliptic. Thus I say that in the course of one day the radius through the centre of the eccentre corresponding to EA rotates uniformly in advance [i.e. in the reverse order] of the signs to the position ED, carrying the apogee of the eccentre to D,[94] and making arc AD 11;9°. [In the same time] the radius through the centre of the epicycle [corresponding to EA] rotates uniformly, again about E, towards the rear through the signs to the position EB, carrying the centre of the epicycle to H, and making arc AB 13;14°. Thus the apparent distance of H, the centre of the epicycle, is 13;14° (in motion of latitude) from the northern limit A, 13;11° (in longitude) from the

---

94   Omitting καί γράφειν περί τό Ζ κέντρον ΔΗ έκκεντρον after Δ at H358,20–21. This would mean 'and describing eccentre DH about centre Z'. This is nonsense: EA does not 'describe the eccentre' (since it is not a radius of the eccentre), but merely marks the position of the apogee of the eccentre. If Ptolemy wanted to refer to the eccentre here, he would presumably have written (as does Is.) καί γράφέντος περί τό Ζ κέντρον τού ΔΗ έκκέντρου 'and if the eccentre DH is described about centre Z'. However, it seems more likely that this is an interpolation by someone who wanted an explicit reference to the drawing of the eccentre DH on centre Z, represented in Fig. 5.1.

beginning of Aries (for the northern limit A has moved to ♓ 29;57° in the same time), and 24;23° (the um of arc AD and arc AB, and twice the mean daily elongation) from the apogee of the eccentre D. Since, in this way, the motion through B and the motion through D meet each other once in half a mean [synodic] month, it is obvious that these motions will always be diametrically opposite at intervals of a quarter and three-quarters of that period, i.e. at the mean quadratures. At those times the centre of the epicycle, located on EB, will be diametrically opposite the apogee of the eccentre, located on ED, and [thus] will be at the perigee of the eccentre.

It is also clear that under these circumstances the eccentre itself (that is, the fact that the arc DB is not similar to arc DH) will not produce any correction to the mean motion. For the uniform motion of the line EB is counted, not along arc DH of the eccentre, but along arc DB of the ecliptic, since it rotates, not about the centre of the eccentre Z, but about E. The only [correction] which will result is that due to the difference in the effect of the epicycle: as the epicycle moves towards the perigee it produces a continuous increase in the equation of anomaly (subtractive and additive alike), since the angle formed by the epicycle at the observer's eye is greater at positions [of the epicycle] nearer the perigee. On the other hand, there will, in general, be no difference from the first hypothesis when the centre of the epicycle is at the apogee A, which is the situation at the mean conjunctions and oppositions.

118

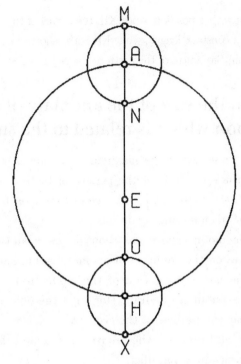

Fig. 5.2

For if [Fig. 5.2][95] we draw epicycle MN about point A,
AE:AM is the same ratio as that which we demonstrated from
the eclipses. The greatest difference will be when the epicycle
reaches H, the perigee of the eccentre (as XO here). This occurs
at the mean quadratures. For the ratio XH:HE is greater than

---

95   The figure given by Heiberg (p. 360), which is taken from the ms. tradition represented
by A, is wrong in making E the centre of the circle and adding a point K above it. My figure
agrees with the text and with part of the Arabic tradition (e.g. P), except that all Arabic mss.
have the equivalent of Θ for O. Manitius already made the same correction, except that he
unnecessarily added the point Z (unattested in the mss.) as the centre of the circle.

that at any other position, since XH, the radius of the epicycle, is always a constant length, while EH is the shortest of all lines drawn from the centre of the earth to the eccentre.

3. On the size of the anomaly of the moon which is related to the sun

In order to see what the maximum equation of anomaly is when the epicycle is at the perigee of the eccentre, we sought observations of the distance of the moon from the sun under the following conditions:

1. The moon's speed was about at the mean (for that is when the equation of anomaly is maximum).
2. The mean elongation of the moon from the sun was about a quadrant (for then the epicycle was near the perigee of the eccentre).
3. In addition to the above, the moon had no longitudinal parallax.

If these conditions are fulfilled, the apparent observed longitudinal distance is the same as the true, and thus we can safely infer the size of the second anomaly which we are seeking. When we investigate on the basis of the above kind of observations, we find that, when the epicycle is closest to the earth, the greatest equation of anomaly is about 7⅔° with respect to the mean position (or 2⅔° different from [the corresponding equation of] the first anomaly).

We will illustrate the way in which this kind of determination is made from one or two observations by way of example. We sighted sun and moon in the 2nd year of Antoninus, Phamenoth [VII] 25 in the Egyptian calendar [139, Feb. 9], after sunrise, and 5¾ equinoctial hours before noon. The sun was sighted in ≈18⅚°, and ♐ 4 was culminating. The apparent position of the moon was ♏ 9⅔° and that was its true position too, since when it is near the beginning of Scorpius, about 1½ hours to the west of the meridian at Alexandria, it has no noticeable parallax in longitude.[96] Now the time from epoch in the first year of Nabonassar to the observation is 885 Egyptian years 203 days is ¾ equinoctial hours (whether reckoned simply or accurately).

For this moment we find:

mean position of the sun ≈ 16;27°

true position of the sun ≈ 18;50° (in accordance with its sighted position according to the astrolabe).[97]

From the first hypothesis we find the mean position of the moon at that moment as ♏ 17;20° (thus its mean elongation from the sun was about a quadrant), and the moon's distance in anomaly from the apogee of the epicycle as 87;19° (which is near the position of maximum

---

96   I.e. at that situation the angle between ecliptic and altitude circle (derived from Table II 13) is about 90°, hence the parallax affects only the latitude, not the longitude. Interpolation in the tables for Clima III, ♏ 9;40°, 1½h west of the meridian, gives 83;5°. Exact computation for Alexandria (φ ≈ 31°) gives 83;45°. For the computations here and at the other observations of V3 and V5 see *HAMA* 91–2.

97   Is this meant as a confirmation of the accuracy of the observation? This would imply that Ptolemy set up the instrument by using the shadow. It may, however, merely mean that this computation is the basis or the position to which Ptolemy set the instrument.

equation). Thus the true position of the moon was less than the mean by 7⅔° (instead of the 5° of the first anomaly).[98]

Again, to display the amount of the equation under similar conditions which is derived from Hipparchus' observations of such positions, we will adduce one of these. He says that he made the observation in the fifty-first year[99] of the Third Kallippic Cycle, Epiphi [XI] 16 in the Egyptian calendar [-127 Aug. 5], when ⅔ of the first hour had passed. 'The speed was [that of day] 241',[100] he says, 'and while the sun was sighted in Leo 8⁷⁄₁₂° the apparent position of the moon was

---

98   Precise computation: mean elongation = ≈ 16;27° − ♏ 17;20° = 89;7°; equation = ♏ 9;40° − ♏ 17;20° = -7;40°; equation from first hypothesis (from Table IV 10), α(87;19°) − -4;57°. However, Ptolemy is operating with rounded numbers, quite properly here.

99   I have, doubtfully, accepted the emendation να' for ν' ('fiftieth') at H363,l 6. The Julian date of the observation, -127 Aug. 5, is guaranteed both by the astronomical data and by Ptolemy's reckoning in the era Nabonassar. Ideler (*Historische Untersuchungen* 217–18) made the emendation because he calculated, correctly, from the known epoch of the Kallippic cycles that this must fall in the fifty-first year. In this case (cf. p. 214 n.72) using the Egyptian calendar makes no difference. However, I suspect that the error, if it is one, lies not with the scribes but with Ptolemy or even Hipparchus, and that possibly there is no error, but another method of counting which eludes us.

100   Literally 'The true daily motion (δρόμος) was the 241st'. Hipparchus is referring to a table of the true motion of the moon over 248 days (≈ 9 anomalistic months), in which the moon was supposed to return to the same velocity. Such a table is extant on a cuneiform tablet, ACT no. 190 (III p. 131). If Hipparchus was using that table the motion on day 241 would be 13;30° or 13;31,10° (according to whether one starts at the beginning or goes in reverse from the end), i.e. close to the mean, as our passage requires. The historical interest of this passage has been missed because '241' has hitherto been interpreted as 'degrees of anomaly' (and hence 'emended', to '259' by Manitius and to μεσος, 'mean', by Halma). I think it likely that Hipparchus was the channel through which use of the 248-day lunar anomaly period was transmitted from Mesopotamia to the Greek world (e.g. Vettius Valens I 4–5, ed. Kroll 20–1, and P. Ryl. 27, on which see *HAMA* 808 ff.), and ultimately to India (the Vākya system, see *HAMA* 817 ff.) See provisionally Toomer [11] p. 108 n.12.

Taurus 12⅓°, and its true position was approximately the same'. So the true observed distance between moon and sun was 86;15°. But when the sun is near the beginning of Leo, at Rhodes (where the observation was made), 1 hour of the day is 17⅓ time-degrees. So the 5⅓ seasonal hours (which make up the interval to [the following] noon) produce 6⅙ equinoctial hours. Therefore the observation occurred 6⅙ equinoctial hours before noon on the sixteenth, while ♉ 9° was culminating. Thus in this case the time from epoch to the observation is

619 Egyptian years 314 days { 17⅚ equinoctial hours reckoned simply
17¾ equinoctial hours reckoned accurately.[100b]

For this moment we find from our hypotheses (since the meridian through Rhodes is the same as that through Alexandria):[101]

mean position of the sun: ♌ 10;27°

true position of the sun: ♌ 8;20°

mean position of the moon in longitude: ♉ 4;25°

(thus the mean elongation was again nearly a quadrant)

100b As Neugebauer remarks, the equation of time for a solar longitude of ♌ 8°should be -16 mins, rather than -5 mins. For this and other inaccuracies in Ptolemy's computations see *HAMA* 92–3.

101 In fact Rhodes is about 1.7° west of Alexandria. The notion that they lay on the same meridian was traditional: see Strabo 2.5.7, where the same meridian is supposed to pass through Meroe, Syene, Alexandria, Rhodes, the Troad, Byzantium and the Borysthenes. This is probably derived from Eratosthenes via Hipparchus.

mean distance of the moon from the apogee of the epicycle in anomaly: 257;47° (which is again near the position of the maximum equation of the anomaly due to the epicycle).

So the distance from the mean moon to the true sun is calculated as 93;55°. And the observed distance from the true moon to the true sun was 86;15°.[102] Therefore the true position of the moon was greater than the mean, again by 7⅔° instead of the 5° of the first hypothesis. And it is [further] evident, that of these two observations taken near the second quadrature, ours was found to be less than the position computed from the first anomaly by 2⅔°, while Hipparchus' was greater by the same amount, since the total equation of anomaly was subtractive at our observation and additive at Hipparchus'.

From numerous other similar observations also we find that the greatest equation of anomaly is about 7⅔° when the epicycle is at the perigee of the eccentre.

4. On the ratio of the eccentricity of the moon's circle

With this as a datum, let [Fig. 5.3] the moon's eccentric circle be ABG on centre D and diameter ADG, on which E is taken as the centre of the ecliptic. Thus A is the apogee of the eccentre and G the perigee. On centre G draw the moon's epicycle ZHΘ, draw EΘB tangent to it, and join GΘ.

---

102   Note that Ptolemy takes only the distance observed by Hipparchus (86;15°) as accurate, and substitutes his own *calculations* of the positions of sun and moon for those observed (or calculated) by Hipparchus.

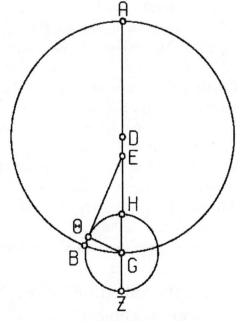

Fig. 5.3

Then since the greatest equation of anomaly occurs when the moon is at the epicycle tangent, and we have shown that this amounts to 7⅔°, the angle at the centre of the ecliptic,

$$\angle \text{ GE}\Theta = \begin{cases} 7;40° \text{ where 4 right angles} = 360° \\ 15;20°° \text{ where 2 right angles} = 360°° \end{cases}$$

Therefore in the circle about right-angled triangle GE⊖
arc G⊖ = 15;20°

125

and the corresponding chord

$G\Theta \approx 16^P$ where the hypotenuse $GE = 120^P$.

So, where $G\Theta$, the radius of the epicycle, is, as was shown, $5;15^P$

and EA, the distance from the centre of the ecliptic to the apogee of the eccentre, is 60^P,

EG, the distance from the centre of the ecliptic to the perigee of the eccentre, is $39;22^P$.

Therefore, by addition, diameter $AG = 99;22^P$, and DA, the radius of the eccentre $= 49;41^P$

and ED, the distance between the centres of the ecliptic and the eccentre $= 10;19^P$

Thus we have demonstrated the ratio of the eccentricity.

5. On the 'direction' of the moon's epicycle [103]

As far as concerns the phenomena at syzygies and at quadrature positions of the moon, the preceding discussion would provide a full explanation of the hypotheses elucidating the circles of the moon described above. But from individual observations taken at distances of the moon [from the sun] when it is sickle-shaped or gibbous (which occur when the epicycle is between the apogee and the perigee of the eccentre), we find that the moon has a peculiar characteristic associated with

---

103   See *HAMA* 88–91, Pedersen 189–95.

the direction[104] in which the epicycle points. Every epicycle must, in general, possess a single, unchanging point defining the position of return of revolution on that epicycle. We call this point the 'mean apogee', and establish it as the beginning from which we count motion on the epicycle. Thus point Z on the previous figure [5.3] is such a point. It is defined, for the position of the epicycle at apogee or perigee of its eccentre, by the straight line drawn through all the centres [of ecliptic, eccentre and epicycle] (DEG here). Now in all other hypotheses [involving epicycle on eccentre], we see absolutely nothing in the phenomena which would count against the following [model]: in other positions of the epicycle [outside apogee and perigee of the eccentre], the diameter of the epicycle through the above apogee, i.e. ZGH, always keeps the same position relative to the straight line which carries the epicycle centre round with uniform motion (here EG), and [thus] (as one would think appropriate) always points towards the centre of revolution, at which, furthermore, equal angles of uniform motion are traversed in equal times. In the case of the moon, however, the phenomena do not allow one to suppose that, for positions of the epicycle between A and G, diameter ZH points towards E, the centre of revolution, and keeps the same position

---

104 πρόσνευσις, used by Neugebauer and Pedersen as a technical term ('prosneusis') for this element of Ptolemy's lunar theory. However, it is hardly that for Ptolemy, as he applies the word in many other contexts.

relative to EG. We do indeed find that the direction
in which [diameter ZH] points is a single, unchanging
point on diameter AG, but that point is neither E, the
centre of the ecliptic, nor D, the centre of the eccentre,
but a point removed from E towards the perigee of the
eccentre by an amount equal to DE. We shall show
that this is so, again, by setting out, from among
the numerous [relevant] observations, two which are
particularly suitable for illustrating our point, since the
epicycle at these observations was at distances halfway
[between apogee and perigee of the eccentre], and the
moon was near apogee or perigee of the epicycle; for in
these situations occur the greatest effects of the above
direction [of the epicycle diameter].

Now Hipparchus records that he observed the sun
and the moon with his instruments[105] in Rhodes in the
197th year from the death of Alexander, Pharmouthi
[VIII] 11 in the Egyptian calendar [-126 May 2], at the
beginning of the second hour. He says that while the
sun was sighted in ♉ 7¾°, the apparent position of the
centre of the moon was ♓ 21⅔°, and its true position
was ♓ 21⅓ + ⅛° [21;27½°].[106] Therefore at the moment

---

105   It is usually assumed that by this is meant an armillary sphere similar to that
described by Ptolemy in V1 (and often, that Hipparchus was the inventor of that
instrument). That may be true, but the vague expression here certainly does not require
it, and whether the data described below do is doubtful. I consider it possible that
Hipparchus used a dioptra of the type described by Heron ('Dioptra', ed. Schone, 187 ff.).

106   On the correction for parallax made by Hipparchus here (which is fairly accurate)
see *HAMA* 92.

in question the distance of the true moon from the true sun was about 313;42°, [counting] towards the rear. Now the observation was made at the beginning of the second hour, about 5 seasonal hours (which correspond to about 5⅔ equinoctial hours in Rhodes on that date) before noon on the 11th. So the time from our epoch to the observation is

620 Egyptian $\begin{cases} 18⅓ \text{ equinoctial hours reckoned simply} \\ 18 \text{ equinoctial hours reckoned accurately.} \end{cases}$
years 219 days

For this moment we find:
mean sun in ♉ 6;41°
true sun in ♉ 7;45°

mean $\begin{cases} \text{in } ♓ 22;13° \text{ in longitude} \\ \text{at } 185;30° \text{ from mean apogee of epicycle in anomaly.} \end{cases}$
moon

Therefore the distance of the mean moon from the true sun was 314;28°.

With these data, let [Fig. 5.4] the moon's eccentric circle be ABG on centre D and diameter ADG, on which E represents the centre of the ecliptic. On centre B draw the moon's epicycle, ZHΘ. Let the sense of motion of the epicycle be towards the rear from B towards A, and the sense of motion of the moon on the epicycle be from Z towards H and [then] Θ. Join DB and EΘBZ.

129

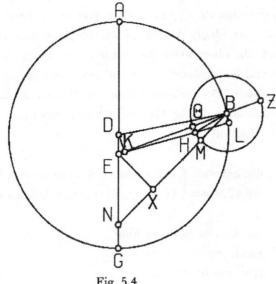

Fig. 5.4

Now in a mean [synodic] month occur two revolutions of the epicycle on the eccentre, and in the situation in question the elongation of mean moon from mean sun was 315;32°. So if we double the latter and subtract [the 360° of] a circle, we will get the elongation at that moment of the epicycle from the apogee of the eccentre, [counting] towards the rear: this is 271;4°.

∴ ∠ AEB = 88;56° (remainder [when 271;4° is subtracted] from 360°).

So drop perpendicular DK from D on to EB.

$$\therefore \angle \text{DEB} = \begin{cases} 88;56° \text{ where 4 right angles} = 360° \\ 177;52°° \text{ where 2 right angles} = 360°° \end{cases}$$

Therefore in the circle about right-angled triangle DEK,

arc DK = 177;52°

and arc EK = 2;8° (supplement).

Therefore the corresponding chords

$$\left. \begin{array}{l} DK = 119;59^P \\ \text{and } EK = 2;14^P \end{array} \right\} \text{ where hypotenuse } DE = 120^P$$

Therefore where DE, the distance between the centres, is $10;19^P$ and DB, the radius of the eccentre, is $49;41^P$,

DK $\approx$ 10; =19^P also, and EK= $0;12^P$.

But $BK^2 = DB^2 - DK^2$

$\therefore$ BK = $48;36^P$ in the same units, and, by addition, BE [=BK+EK] = $48;48^P$.

Again, since the distance of the mean moon from the true sun was found to be 314;28°, and the distance of the true moon (from the true sun) was observed to be 313;42°, the equation of anomaly is -0;46°. Now the mean position of the moon is seen along the line EB. So let the moon be located at H (since it is near the perigee), join EH and BH, and drop perpendicular BL from B on to EH produced. Then, since ∠ BEL contains the moon's equation of anomaly,

$$\angle \text{BEL} = \begin{cases} 0;46° \text{ where 4 right angles} = 360° \\ 1;32°° \text{ where 2 right angles} = 360°° \end{cases}$$

Therefore in the circle about right-angled triangle EBL,

arc BL = 1;32°

131

and the corresponding chord

$BL = 1;36^P$ where the hypotenuse $EB = 120^P$.

Therefore where $BE = 48;48^P$ and BH, the radius of the epicycle, is $5;15^P$,

$$BL = 0;39^P.$$

Therefore where BH, the radius of the epicycle, is 120^P,

$$BL = 14;52^P$$

and, in the circle about right-angled triangle BHL,

$$\text{arc } BL = 14;14°$$

∴ ∠ BHL = $14;14°°$ where 2 right angles = $360°°$,

and, by subtraction [of ∠ BEL], ∠ EBH $= \begin{cases} 12;42°° \text{ where 2 right angles} = 360°° \\ 6;21° \text{ where 4 right angles} = 360° \end{cases}$

That [6;21°], then, is the size of arc HΘ of the epicycle, which comprises the distance from the moon to the true perigee [of the epicycle].

But since the distance of the moon from the mean apogee at the time of the observation was $185;30°$ [p. 228], it is clear that the mean perigee is in advance of the moon, i.e. of point H. Let [the mean perigee] be point M, draw line BMN, and drop perpendicular EX on to it from point E.

Then since, as was shown,

$$\text{arc } ΘH = 6;21°,$$

and arc HM, the distance from the perigee, is given as $5;30°$, by addition, arc $ΘM = 11;51°$.

So ∠ EBX= $\begin{cases} 11;51° \text{ where 4 right angles}= 360° \\ 23;42°° \text{ where 2 right angles}= 360°° \end{cases}$

Therefore in the circle about right-angled triangle BEX,

arc EX = 23;42°

and EX = 24;39^P where hypotenuse BE = 120^P.

Therefore where BE = 48;48^P

EX = 10;2^P.

Again, since

∠ AEB = 177;52°° and ∠ EBN = 23;42°° $\Big\}$ where 2 right angles = 360°,

by subtraction, ∠ ENB = 154;10°°.

Therefore in the circle about right-angled triangle ENX,

arc EX= 154;10°

and EX = 116;58^P where hypotenuse EN = 120^P.

Therefore where EX = 10;2^P and DE, the distance

between the centres, is 10;19^P,

EN = 10;18^P.

Therefore the [radius of the epicycle] through the mean perigee, BM, points in a direction such that, when produced to N, it cuts off a line EN which is very nearly equal to DE.

Similarly, in order to show that we get the same result at the opposite sides of eccentre and epicycle, we have again selected from the distances [between sun and moon] observed by Hipparchus, as already mentioned, in Rhodes, the observation he made in the same year [as the preceding one], being the 197th year from the death of Alexander, Payni [X] 17 in the Egyptian calendar [-126 July 7], at 9⅓ hours. He says that while the sun was sighted at ♋ 10⁹/₁₀° the apparent position of the moon was ♌ 29°. And this was its true position too; for at Rhodes, near the end of Leo, about one hour past the meridian, the moon has no longitudinal parallax.[107] Therefore the distance of the true moon from the true sun at the time in question was 48;6° towards the rear. Now since the observation was 3⅓ seasonal hours after noon on the 17th of Payni, which correspond to about 4 equinoctial hours in Rhodes on that date, the time from our epoch to the observation is

620 Egyptian years 286 days
{ 4 equinoctial hours reckoned simply
3⅔ equinoctial hours reckoned accurately.

For this moment we find:
mean sun at ♋ 12;5°
true sun at ♋ 10;40°
mean moon at ♌ 27;20° in longitude

(thus the distance of the mean moon from the true sun was 46;40°) mean moon at 333;12° in anomaly from the apogee of the epicycle.[108]

107 For verification of this see *HAMA* 92.

108 For 620y 286d 3⅔h I find: $\bar{\lambda}$ ☽ = 147;7°, $\bar{\alpha}$ ☽ = 333;1°. Since the differences from Ptolemy's positions represent the lunar motion over about 20 mins., it is obvious that he has

With these data, let [Fig. 5.5] the moon's eccentric circle be ABG on centre D

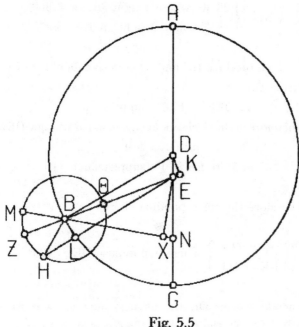

Fig. 5.5

and diameter ADG, on which the centre of the ecliptic is represented by point E. About point B draw the moon's epicycle, ZHΘ, and join DB, EΘBZ.

___

carelessly calculated the positions for 4 hours after noon, i.e. without making the correction for the equation of time, which he had given, correctly, as about 20 mins. This error has a not inconsiderable effect on the final result, which would not agree nearly so neatly if the computation were carried out with the above figures.

Then since twice the mean elongation of sun and moon is 90;30°, from the theory already established

$$\angle\ AEB = \begin{cases} 90;30° \text{ where 4 right angles} = 360° \\ 181°° \text{ where 2 right angles} = 360°° \end{cases}$$

So if we produce BE and drop perpendicular DK on to it from D,

$$\angle\ DEK = 179°° \text{ (supplement).}$$

Therefore in the circle about right-angled triangle DEK

arc DK= 179°

and arc EK = 1° (supplement).

Therefore the corresponding chords

DK= 119;59^p
and EK = 1;3^p } where hypotenuse DE = 120^p

Therefore where DE, the distance between the centres, is 10;19^p and BD, the radius of the eccentre, is 49;41^p,

$$DK \approx 10;19^p$$
$$\text{and } EK = 0;5^p.$$

Now since $BK^2 = BD^2 - DK2$,

$$BK = 48;36^p,$$

and, by subtraction [of EK], EB= 48;31^p.

Furthermore, since the distance of mean moon from

true sun was found to be 46;40°, and the distance of true moon [from true sun was observed as] 48;6°, the equation of anomaly is +1;26°. So let the position of the moon be at H (since it is near the apogee of the epicycle). Join EH, BH, and drop perpendicular BL from B on to EH.

Then since

$$\angle \text{ BEL} = \begin{cases} 1;26° \text{ where 4 right angles} = 360° \\ 2;52°° \text{ where 2 right angles} = 360°°, \end{cases}$$

in the circle about right-angled triangle BEL,

arc BL = 2;52°

and BL= 2;59ᴾ where hypotenuse EB = 120ᴾ.

Therefore where EB = 48;31ᴾ and BH, the radius of the epicycle, is 5;15ᴾ,

BL= 1;12ᴾ.

So in the circle about right-angled triangle BHL,

BL = 27;34ᴾ where hypotenuse BH = 120p,[109]

and arc BL = 26;34°.

∴ ∠ BHL = 26;34°° where 2 right angles = 360°°, and, by addition [of ∠ BEL = 2;52°°],

$$\angle \text{ ZBH} = \begin{cases} 29;26°° \text{ where 2 right angles} = 360°° \\ 14;43° \text{ where 4 right angles} = 360°. \end{cases}$$

---

109   1;12 x 120/5; 15 = 27;25,43. Ptolemy was obviously operating, not with the value 1;12, but with I; 12,22 (which leads to 27;34,5), which is in fact what one finds from the immediately preceding calculation, 2;59 x 48;31/120.

That [14;43°] is the size of the arc HZ of the epicycle, which comprises the distance from the moon to the true apogee.

But since [the moon's] distance from the mean apogee at the time of the observation was 333;12°, if we put the mean apogee at M, draw line MBN, and drop perpendicular EX on to it from E, then

arc HZM = 26;48° (by subtraction [of 333;12°] from the circle), and, by subtraction [of arc HZ = 14;43°], arc ZM = 12;5°.

$$\therefore \angle \text{MBZ} = \angle \text{EBX} = \begin{cases} 12;5° \text{ where 4 right angles} = 360° \\ 24;10°° \text{ where 2 right angles} = 360°°. \end{cases}$$

Therefore in the circle about right-angled triangle BEX
$$\text{arc EX} = 24;10°$$
and EX= 25;7^P where hypotenuse BE = 120^P.

Therefore where BE = 48;31^P and DE, the line between the centres, is 10;19^P,

$$\text{EX} = 10;8^P.$$

Again, since $\angle$ AEB is given as 181°° where 2 right angles = 360°°, and we have shown that $\angle$ EBN = 24;10°°, by subtraction, $\angle$ ENB = 156;50°° in the same units, and, in the circle about right-angled triangle ENX,
$$\text{arc EX} = 156;50°$$
and EX = 117;33^P where hypotenuse EN = 120^P.

Therefore where EX= 10;8^p and DE, the line between the centres, is 10;19^P,

$$\text{EN} = 10;20^P.$$

So from this calculation too it turns out that MB, [the radius of the epicycle] through M, the mean apogee, points in a direction such that, when produced to N, it cuts off a line EN approximately equal to DE, the distance between the centres.

We also find that approximately the same ratio results by calculation from a number of other observations. Thus these observations confirm the peculiar characteristic of the direction of the epicycle in the hypothesis of the moon: the [uniform] revolution of the centre of the epicycle takes place about E, the centre of the ecliptic, but the diameter of the epicycle which defines the unchanging point of the epicycle at which the mean epicyclic apogee is located points, not (as it does for the other [planets]), towards E, the centre of mean motion, but always towards N, which is removed in the opposite direction [to D from E] by an amount equal to DE, the distance between the centres.

11. On the moon's parallaxes[110]

With the above we have about disposed of the [elements] necessary for finding the true positions of the moon. However, in the case of the moon there is the additional problem that its apparent position does not coincide with its true position, even to the senses. For, as we said, the earth does not bear the ratio of a point to the distance of the moon's sphere.

---

110   On chs. 11 and 12 see *HAMA* 100–1, Pedersen 203–4.

Hence it is both necessary and appropriate to discuss the lunar parallaxes, especially in order to deal with the theory of solar eclipses, amongst other phenomena. By means of the lunar parallaxes it will be possible, given a true position [of the moon], [i.e. its position] with respect to the centre of the earth and of the ecliptic, to determine its position as seen from the standpoint of the observer, that is from some point on the earth's surface, and, *vice versa*, to determine the true position from the apparent position. Now it is a feature of this kind of enquiry that one cannot find the amount of the parallax for individual situations unless one is first given the ratio of the distance [of the body to the earth's radius], nor can one find the ratio of the distance without the parallax for some particular situation being given. Hence for those bodies with no perceptible parallax, namely, those to [the distance of] which the earth bears the ratio of a point, it is, obviously, impossible to find the ratio of the distance. But in the case of those bodies, like the moon, which do exhibit a parallax, the only appropriate procedure is, first given some particular parallax, to find the ratio of the distance. For it is possible to make an observation of a [particular] parallax of this kind by itself, but quite impossible to determine the amount of the distance [by itself].

Now Hipparchus used the sun as the main basis of his examination of this problem. For since it follows from certain other characteristics of the sun and moon (which we shall discuss subsequently) that, given the distance to one of the luminaries, the distance to the other is also given, Hipparchus tries to demonstrate the moon's distance by guessing at the sun's. First he supposes that the sun has the least perceptible parallax, in order to find its distance, and then he uses the solar eclipse which he adduces; at one time he assumes that the sun has no perceptible parallax, at another that it has a parallax big enough [to be observed]. As a result the ratio of the moon's distance came out different for him for each of the hypotheses he put forward; for it is altogether uncertain in the case of the sun, not only how great its parallax is, but even whether it has any parallax at all.[111]

12. On the construction of a parallactic instrument[112]

We, in contrast, to avoid taking any uncertain factors into our examination of this topic, constructed an instrument to enable us to observe as accurately as possible the amount of the moon's parallax, and its zenith distance, along the great circle through the poles of the horizon and the moon.

---

111 This passage is supplemented by Pappus' commentary ad loc. (Rome[l] I 67–8), which extracts some details of the two procedures of Hipparchus from Books 1 and 2 respectively of the latter's 'On sizes and distances'. For details of the important historical consequences which can be drawn see Toomer[9] (showing that the solar eclipse referred to is that of -189 Mar. 14), which builds on the work of Swerdlow, 'Hipparchus'.

112 On the instrument described in this chapter (known in the middle ages as a 'triquetrum' see Price, 'Precision Instruments' 589–90 with Fig. 344. My Fig. G is based on the text of the Almagest rather than on the figure provided by Pappus in his commentary (Rome[1] I p. 71, with a modern reconstruction; see also Rome's notes on pp. 70–5).

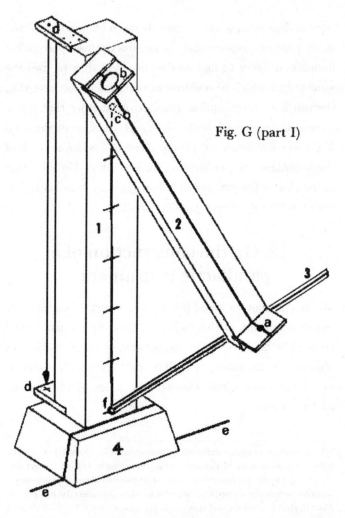

Fig. G (part I)

We made two rods [Fig. G,1,2], rectangular [in cross-section], no less than 4 cubits long, so as to admit finer graduation, and with a cross-section of sufficient size that

they were not distorted because of their length, but each
side conformed very strictly to a straight line. Then we
drew a straight line along the middle of the broader side
of each rod, and affixed to one of them [Fig. G,2], at
each end, centred on the line, and perpendicular [to it],
two rectangular plates, of equal size and parallel to each
other [Fig. G,a,b]; each plate had an aperture exactly
in the centre, the aperture at the eye being small, and
that towards the moon being greater, in such a way that
when one eye was placed at the plate with the smaller
aperture, the whole of the moon would be visible through
the aperture on the other plate, which was aligned [with
the first aperture]. We made a perforation of equal size
through both rods at the end of the median line near
the plate with the larger hole, and fitted a peg [Fig. G,c]
through both perforations in such a way that the sides of
the rods inscribed with the lines[113] were fastened together
round the peg as a centre, but the rod with the plates
could rotate freely in all directions without distortion.
We wedged the rod with no plates on it [Fig. G,l] into a
base [Fig. G,4]. On the median line of each rod, at the
end by the base, we took a point as far as possible from
the centre of the peg (the same distance from it [on both
rods]), and, on the rod with the base, divided the line
so defined into 60 sections, subdividing each section into

---

113   The faces of rods 1 and 2 inscribed with the lines cannot be flush with one another,
as is clear from Fig. G. Ptolemy seems to mean only that one views the inscribed faces of
the two rods as radii of a circle with centre peg c.

as many subdivisions as possible. We also attached to
the back of the same rod, at its end, [two] plates [Fig.
G,d,d] having their corresponding faces aligned with each
other,[114] and each being equidistant in all respects from
that same median line, so that when a plumb-line was
suspended between them, the rod could be set up exactly
perpendicular to the plane of the horizon. We also had a
meridian line [Fig. G,e] ready drawn in the plane parallel
to that of the horizon in an unshaded place.

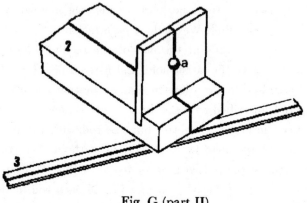

Fig. G (part II)

---

114 Excising the words πρός τῇ αὐτῇ γραμμῇ at H404, 17–18. That would mean 'each
having that face which was on the same side as the [graduated] line aligned with the other'.
But this is impossible, since the plates are not to one side of the face with the graduations,
but 'on the back', i.e. on the face opposite the graduated line. This is also clear from Pappus'
detailed description (Rome p.75). πρός τῇ γραμμῇ is a stupid gloss on ἐπί τά αὐτά μέρη,
which I have translated 'corresponding', but which literally means 'in the same direction'.
The interpolation is old, since it is found in the Arabic tradition.

We set the instrument upright in such a way that the sides of the rods which were held flush with each other by the peg lay in the meridian, being parallel to the meridian line, and the rod with the base was fixed exactly perpendicular, in a firm and immovable position, while the other rod could move in the plane of the meridian about the peg, responding to the pressure [of the user].[115] We also added another thin, straight rod, [Fig. G,3] attached by a small pin [Fig. G,f] at the base end of the graduated line, so that it too could be rotated, and long enough to reach the end of the line on the other rod equidistant [from the peg] when it was rotated to its maximum distance [from the base];[116] thus by rotating it at the same time as the latter, one could use it to show the straight-line distance between the ends [of the centre-lines on the two rods].

We made our observations of the moon as follows. The moon had to be located on the meridian, and near

---

115   i.e. the peg held the rods together tightly enough so that rod 2 would not move under its own weight, but loosely enough so that it could be rotated by the user.

116   This rod has indeed to be 'thin', since it has to pass between the two rods 1 and 2, the faces of which are supposed to be flush. Pappus overcomes this difficulty by saying that rod 2 has to be hollowed out along its length to the depth of the thickness of rod 3 (Rome p. 73). There is the further difficulty that according to Ptolemy's instructions rod 3 has to be long enough to reach to the end of rod 2 at the maximum rotation, presumably 90°: hence its length should be ($\sqrt{2}$ x length of the graduated line). But since one measures the chord of the zenith distance, not directly on rod 3, but by marking it on rod 3 and then measuring it on the scale on rod 1, no zenith distance greater than 60° (the chord of which is 60^p) can be measured. Hence, presumably, Pappus (p. 73) says that rod 3 should be less than the length of the graduated line. Rome (p. 73 n.0) suggests that Ptolemy deliberately chose this limit to avoid the complications of refraction near the horizon. It seems more likely that it is simply a by-product of Ptolemy's construction, and that Pappus' shortening of the rod was done to avoid the difficulties which would result from trying to apply rod 3 to the graduated line if it were 60^p or more.

the solstices on the ecliptic, since at such situations the great circle through the poles of the horizon and the centre of the moon very nearly coincides with the great circle through the poles of the ecliptic, along which the moon's latitude is taken. Furthermore the true distance [of the moon] from the zenith can also be conveniently determined from the same situation. When the moon was precisely in the meridian, we moved the rod with the [sighting-] plates on it round to the position in which the centre of the moon, when sighted through both apertures, was in the centre of the larger aperture. We marked on the thin rod the distance between the ends of the lines on the [two] rods, then applied the distance [marked on the thin rod] to the line on the upright rod graduated into 60 sections. Thus we found the amount of that distance in those units of which the radius of the circle described by the rotation [of the rod with the sighting-plates] in the plane of the meridian contains 60. By calculating the arc corresponding to that chord, we found the angular distance of the apparent centre of the moon from the zenith, measured along the great circle through the poles of the horizon and the moon's centre, which coincided at that moment with the [great circle] through the poles of the equator and the ecliptic, [i.e.] the meridian.

In order, first, to determine the precise amount of the moon's greatest deviation in latitude, we made sightings when the moon was simultaneously near the summer solstice and near the northern limit of its inclined circle.[117] For in the region of those points the moon's latitude remains sensibly the same over a considerable interval, and furthermore, since the moon is then very near the zenith at the parallel through Alexandria (at which we made our observations), its apparent position is approximately the same as its true position. At such situations it was found that the distance of the centre of the moon from the zenith was always about 2⅛°. Hence by this method too the moon's greatest latitude either side of the ecliptic is shown to be 5°. For the zenith distance of the equator at Alexandria has been shown to be 30;58°; if we subtract from this the 2⅛° (which is the apparent distance [of the centre of the moon from the zenith]), the result [28;50½°] is about 5° greater than the distance from the equator to the summer solstice, which was shown to be 23;51°.

Then, in order to attack the problem of the parallaxes, we observed the moon in the same way, but this time when it was near the winter solstice, both for the reason already mentioned [above] and because its distance from the zenith in that situation is the greatest of all such meridian positions, and thus provides us with a greater and more easily determinable parallax. We will set out

---

117 Since the revolution of the node takes place once in about 18⅔ years, this situation occurs 9⅓ years earlier or later than the similar situation of the moon near the winter solstice, observed by Ptolemy (V 1 3) in Oct. 135. Therefore these observations were made either in the summer of 126, or in the spring of 145. This is the only useful conclusion that can be drawn from the confused discussion of Newton, 184–6.

one of a number of parallax observations which we made at such situations. By this means we shall display the method of calculation and at the same time provide a demonstration of the rest of what is to follow in the appropriate order.

13. Demonstration of the distances of the moon[118]

In the twentieth year of Hadrian, Athyr [III] 13 in the Egyptian calendar [135 Oct. 1], 5⅚ equinoctial hours after noon, just before sunset, we observed the moon when it was on the meridian. The apparent distance of its centre from the zenith, according to the instrument, was $50^{11}/_{12}°$. For the distance [measured] on the thin rod was $51^{7}/_{12}$ of the 60 subdivisions into which the radius of revolution had been divided, and a chord of that size subtends an arc of $50^{11}/_{12}°$. Now the time from epoch in the first year of Nabonassar to the moment of the above observation is

882 Egyptian $\left\{ \begin{array}{l} \text{5⅚ equinoctial hours reckoned simply} \\ \text{5⅓ equinoctial hours reckoned accurately.} \end{array} \right.$
years 72 days

For this moment we find:
mean longitude of the sun: ♎ 7;31°

---

118 See *HAMA* 101–3, Pedersen 204–7.

true longitude of the sun: ♎ 5;28°
mean longitude of the moon: ♐ 25;44°
elongation: 78;13°
distance [in anomaly]
from mean apogee of epicycle: 262;20°
distance in [argument of]
latitude from the northern limit: 354;40°.

Hence the complete equation of anomaly, derived
from the appropriate table, was +7;26°, so that
the true position of the moon at that moment was:
in longitude: ♑ 3;10°
in [argument of] latitude
on the inclined circle: 2;6° from the northern limit
in latitude on the great circle
through the poles of the ecliptic
(which almost coincided at that moment with the
meridian):[119] 4;59° north of the ecliptic.

Now ♑ 3;10° is 23;49° south of the equator on the same
[meridian] circle, and the equator is, likewise, 30;58° south of the
zenith at Alexandria. Therefore the true distance of the centre
of the moon from the zenith was [23;49 + 30;58 – 4;59 =] 49;48°.
And its apparent distance was 50;55°. Therefore the moon's
parallax at the distance [of the moon from the earth]
corresponding to the position in question was 1;7° along the

---

119   For the moon was almost at the winter solstice.

great circle through the moon and the poles of the horizon, when its true distance from the zenith was 49;48°.

Now that we have established that, draw [Fig. 5. 10] in the plane of the great circle through the poles of the horizon and the moon the following great circles, on the same centre:

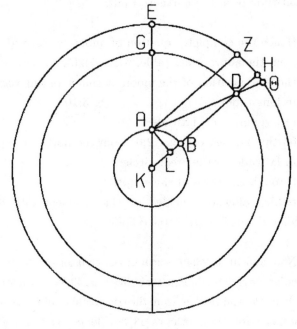

Fig. 5.10

that of the earth, AB; that through the centre of the moon at the [above] observation, GD; the great circle to which the earth bears the ratio of a point, EZHΘ.

Let their common centre be K, and let the line through the points at the zenith be KAGE. Let us assume that the same distance of the moon, D, from the zenith at G is the amount already determined, 49;48°. Join KDH, ADΘ, and furthermore from point A, which represents the observer's eye, draw AL as perpendicular to KB, and AZ as parallel to KH.

Then it is obvious that for an observer at point A the moon's parallax was arc HΘ. So arc HΘ is 1;7°, according to the calculation from the observation. But since arc ZΘ is negligibly greater than arc HΘ (for the whole earth bears the ratio of a point to circle EZHΘ), arc ZHΘ is very nearly the same, 1;7°. And since, again, point A is negligibly different from the centre of circle ZΘ,

$$\angle \; ZA\Theta = \begin{cases} 1;7° \text{ where 4 right angles} = 360° \\ 2;14°° \text{ where 2 right angles} = 360°°. \end{cases}$$

And $\angle$ ADL = $\angle$ ZAΘ = 2;14°°.
Therefore in the circle about right-angled triangle ADL,
$$\text{arc AL} = 2;14°$$
and Crd arc AL = 2;21^P where hypotenuse AD
$$= 120^P.$$
But LD is negligibly smaller than AD.
Therefore where LA = 2;21^P, LD ≈ 120^P.
Furthermore since, by hypothesis, arc GD = 49;48°, the angle at the centre of the circle,

151

$$\angle \text{ GKD} = \begin{cases} 49;48° \text{ where 4 right angles}= 360° \\ 99;36°° \text{ where 2 right angles} = 360°°. \end{cases}$$

Therefore in the circle about right-angled triangle ALK
arc AL = 99;36°
and arc LK = 80;24° (supplement).
Therefore the corresponding chords

$$\left. \begin{array}{l} \text{AL} = 91;39^P \\ \text{and LK=77; 27}^P \end{array} \right\} \text{ where hypotenuse AK=120}^P$$

Therefore where AK, the radius of the earth, is 1^P
AL= $0;46^P$
and KL = $0;39^P$.
But where AL = $2;21^P$, LD, as was shown,= 120^P.
Therefore where AL = $0;46^P$, LD = $39;6^P$.
And, in the same units, KL = $0;39^P$.
and the radius of the earth, KA = 1^P.
Therefore where KA, the radius of the earth, is 1^P,
by addition, KLD, which represents the distance of
the moon at the observation, is $39;45^P$.[120]

Now that we have demonstrated this, let [Fig. 5.11]
the moon's eccentre be ABG on centre D and diameter
ADG, on which E is taken as the centre of the ecliptic,
and Z as the point towards which [the mean apogee
diameter of] the epicycle is directed. Draw the epicycle,

---

[120] There is an accumulated error here, due to a series of small inaccuracies and
roundings. More accurate would be $39;50^P$.

HΘKL, on point B, and join HBΘE, BD and BKZ. Let L represent the position of the moon at the observation in question, and draw perpendiculars to BE, DM from D[121] and ZN from Z. Then since the amount of the elongation at the time of the observation was 78;13°, it follows from the theory previously established that

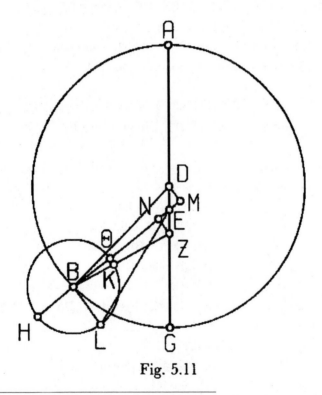

Fig. 5.11

---

121   Heiberg rightly excised ἐκβληθεῖσαν ('extended') at H413,7 as an unnecessary gloss which disturbs the sentence structure. Transferring it after BE (as Halma and Manitius) is no improvement, since the perpendicular from Z is not on the extension of BE.

∠ AEB = 156;26° where 4 right angles= 360°;

hence its supplement, ⎰ 23;34° where 4 right angles = 360°
∠ ZEN = ∠ DEM = ⎱ 47;8°° where 2 right angles = 360°°

Therefore in the circles about the corresponding right-angled triangles, [ZEN, DEM], since DE = EZ,
arc DM = arc ZN = 47;8°
and arc EM = arc EN = 132;52° [supplements].
Therefore the corresponding chords

DM = ZN = 47;59^P ⎱ where hypotenuse DE =
and EM = EN 110;0^P ⎰ hypotenuse EZ = 120^P

Therefore where DE = EZ = 10;19^P and DB, the radius of the eccentre, is 49;41^P,
DM = ZN = 4;8^P
and EM = EN = 9;27^P.
And since $BM^2 = BD^2 - DM^2$,
BM= 49;31^P.
And BE = [BM − EM =] 40;4^P,
and, by subtraction [of EN from BE], BN = 30;37^p where ZN = 4;8^P.
And since $BN^2 + ZN^2 = BZ^2$,
hypotenuse BZ = 30;54^P.

Therefore in the circle about right-angled triangle BZN, where hypotenuse BZ = 120^P

$$ZN = 16;2^P$$
and arc ZN= 15;21°.

$$\therefore \angle \text{ ZBN} = \begin{cases} 15;21°° \text{ where 2 right angles} = 360°° \\ \text{about } 7;40° \text{ where 4 right angles} = 360°. \end{cases}$$

That [7;40°], then, is the size of arc ΘK of the epicycle.

Next, the distance of the moon from the mean apogee of the epicycle at the moment of the observation was 262;20°, and, obviously, its distance from K, the mean perigee, was 82;20° (by subtraction of a semi-circle).

Therefore arc KL = 82;20°

and arc ΘKL = [arc ΘK + arc KL =] 90;0°.

So ∠ ΘBL is a right angle.

$$\therefore EL^2 = BL^2 + EB^2,$$

and where DB, the radius of the eccentre, is 49;41ᴾ

and BL, the radius of the epicycle, is 5;15ᴾ,

EB, as we showed = 40;4ᴾ.

$$\therefore EL = 40;25^P.$$

Therefore the distance of the moon at the observation is 40;25ᴾ,

where BL, the radius of the epicycle, is 5;15ᴾ

and where EA, the distance from the centre of the earth to the apogee of the eccentre, is 60ᴾ,

and where EG, the distance from the centre of the earth to the perigee of the eccentre, is 39;22ᴾ.

But we showed that the moon's distance at the observation, that is EL, was 39;45ᴾ where the radius of the earth is 1ᴾ.

155

Therefore where EL, the distance of the moon at the observation, is 39;45^P, and the earth's radius is 1^P,

EA, the mean distance at the syzygies = 59;0^P,[122]

EG, the mean distance at the quadratures = 38;43^P, and the radius of the epicycle = 5;10^P.

Q.E.D.

14. On the ratio of the apparent diameters of sun, moon and shadow at the syzygies[123]

Now that we have demonstrated the distances of the moon in the above manner, the appropriate sequel is to demonstrate those of the sun as well. This too can readily be performed geometrically, if we are given, in addition to the distances of the moon at the syzygies, the sizes of the angles formed at the [observer's] eye at the syzygies by the diameters of the sun, moon and shadow.

Of the various methods used to solve the latter problem, we have rejected those claiming to measure

---

122 This result for the moon's mean distance agrees well with the facts (it is slightly greater than 60 earth-radii), which means that Ptolemy's parallax at syzygies (i.e. at solar eclipses) is fairly accurate. However, the process by which it is reached contains a number of errors (in the observed parallax, the latitude, the declination etc., and in the distance resulting from Ptolemy's model), which 'miraculously' cancel each other out. For details see *HAMA* 102–3. This is no accident: Ptolemy knew (approximately) what the parallax had to be at eclipses and chose an observation which produced that amount. For a suggestion that the figure of 59 earth-radii had already been derived by Hipparchus see Toomer[9] 131.

123 The chapter heading is placed by most Greek mss. (and by Heiberg's text) before H416, 20. I have transferred it here (before H416,9), following the Arabic mss. (cf. also D, which has it in the upper margin), as a more appropriate break. On ch. 14 see *HAMA* 103–8, Pedersen 207–9 (with the corrections Toomer [3] 140, 143, 149).

the luminaries by measuring [the flow of] water or by the time [the bodies] take to rise at the equinox,[124] since such methods cannot provide an accurate result for the matter in hand. Instead, we too constructed the kind of dioptra which Hipparchus described, which uses a four-cubit rod,[125] and, observing with this, found that the sun's diameter always subtends approximately the same angle, there being no noticeable difference due to [the variation in] its distance, but that the moon subtends the same angle as the sun only when it is at its greatest distance from the earth (i.e. the apogee of the epicycle) at full moon, in contradiction to the hypotheses of my predecessors, [who assumed that it subtends the same angle as the sun at full moon] when it is at mean distance.[126] Furthermore, we find that the angles themselves are considerably smaller than those

---

124   According to Pappus ad loc. (Rome[I] I 87–9) 'the more ancient astronomers' used water clocks to measure the time taken by the sun to cross the horizon, a procedure criticised by Hipparchus. He refers to a lost work of Heron, περί ὑδρίων ὡροσκοπείων, on which see also Proclus, Hypotyposis IV 73–6 (ed. Manitius p. 120–2). At H416,21 Heiberg rightly accents ὑδρομετριῶν (from the abstract ὑδρομετρία). There is no evidence for the existence of ὑδρομετρίον, 'vessel for measuring flow of water', conjectured by LSJ s.v. In the corresponding passage Proclus p. 120 line 14 we should read ὑδρολογίων also *HAMA* 103 n. 1.

125   There are ancient descriptions of this instrument by Pappus in his commentary ad loc. (Rome[I] I 90–2) and by Proclus, Hypotyposis IV 87–96 (ed. Manitius pp. 126–30). See Price, 'Precision Instruments' 591, and, for modern literature, *HAMA* 103 n.2. The essential feature is a plate (πρισμάτιον H417, 22–3) which can be moved along a graduated rod until it appears to exactly cover the object being sighted by the eye placed at one end of the rod.

126   It was shown by Swerdlow, 'Hipparchus' 291–8, that Hipparchus was one of those who held this. An important consequence of this hypothesis is that annular solar eclipses become possible, whereas under Ptolemy's assumption they are impossible.

traditionally accepted.[127] However our computation of the latter rests, not on measurement with the dioptra, but on certain lunar eclipses. For although it was possible to determine readily from the dioptra, as constructed, when both diameters subtend the same angle (since such a determination involves no actual measurement), the *amount* [of the angle subtended] seemed utterly dubious to us, since the measurement[128] involving the positioning of the width [of the plate] which covers [the body being sighted] on the length of the rod running from the eye to the plate can be inaccurate. However, once it was determined that the moon is at its greatest distance when it subtends the same angle at the eye as the sun, we computed the size of the angle it subtends from observations of lunar eclipses in which the moon was near that [greatest] distance, and thence obtained immediately the size of the angle subtended by the sun.

---

127 Hipparchus (see IV 9 p. 205) assumed that the moon at mean distance subtends a six hundred and fiftieth of its circle, or about 0;33,14°; hence his figure for the sun's diameter was the same. Ptolemy (below) finds that when moon and sun have the same apparent diameter (at maximum distance) it is 0;31,20°, considerably smaller. This must be what he means here. However, his value for the lunar diameter at *mean* distance, 0;33,20°, is negligibly different from Hipparchus'.

128 Excising πλείστης οὔσης; at H417,23, to which I can attach no meaning (it cannot mean 'very laborious', as Manitius translates, nor, if it could, would it be true). The variant πλείσταις οὔσαις found in D, part of the Arabic tradition (L) and Pappus (Rome [I] I 93,21) can be translated ('involving multiple positionings'), but it is not true that sighting the moon would require more than one positioning of the plate. Unless the corruption lies deeper (e.g. πλείστης; has replaced a word meaning 'delicate') one must assume that πλείστης οὔσαις was an inept gloss intended to explain why the process was inaccurate, and that this was corrupted to the unintelligible πλείστης οὔσης by attraction to παραμετρήσεως.

We shall explain the method of procedure in this by means of two of the eclipses used.

In the fifth year of Nabopolassar, which is the 127th year from Nabonassar, Athyr [III] 27/28 in the Egyptian calendar [-620 Apr. 21/22], at the end of the eleventh hour in Babylon, the moon began to be eclipsed; the maximum obscuration was ¼ of the diameter from the south. Now, since the beginning of the eclipse occurred 5 seasonal hours after midnight, and mid-eclipse about 6 [seasonal hours after midnight], which correspond to 5⅚ equinoctial hours at Babylon on that date (for the true position of the sun was ♈ 27;3°), it is clear that mid-eclipse, which is when the greatest part of the diameter is immersed in the shadow, occurred 5⅚ equinoctial hours after midnight in Babylon, and exactly 5 [hours after midnight] at Alexandria.[129]

The time from epoch is

126 Egyptian $\left\{\begin{array}{l} \text{17 equinoctial hours reckoned simply} \\ \text{16¾ equinoctial hours in mean solar days}^{129b} \end{array}\right.$
years 86 days

Therefore the lunar position was as follows: mean position in longitude: ♎ 25;32°

true position in longitude: ♎ 27;5°

129 Oppolzer no. 90I: mid-eclipse 2;38 a.m. (≈4½ h after midnight at Alexandria), magnitude 1.6d. P.V. Neugebauer, *Spezieller Kanon*, gives about 5¼h after midnight (Babylon) for mid-eclipse, magnitude 2. 1ᵈ.

129b The equation of time for a solar longitude of ♈ 27° is about -20 mins. rather than -15 mins.

distance [in anomaly] from the apogee of the epicycle: 340;7°

distance [in latitude] from the northern limit on the inclined circle: 80;40°.

Thus it is clear that when the centre of the moon near its greatest distance is 9⅓° distant from the node, measured along its inclined circle, and the centre of the shadow lies on the great circle drawn through the moon's centre at right angles to the inclined circle (which is the situation at which the greatest obscuration occurs), ¼ of the moon's diameter is immersed in the shadow.

Again, in the seventh year of Kambyses, which is the 225th year from Nabonassar, Phamenoth [VII] 17/18 in the Egyptian calendar [-522 July 16/17], 1 [equinoctial] hour before midnight at Babylon, the moon was eclipsed half its diameter from the north. Thus this eclipse occurred about 15/6 equinoctial hours before midnight at Alexandria.[130] The time from epoch is

224 Egyptian years 196 days $\begin{cases} 10\frac{1}{6} \text{ equinoctial hours reckoned simply} \\ 9\frac{5}{6} \text{ equinoctial hours reckoned accurately} \end{cases}$

130 Oppolzer no. 1056: mid-eclipse 2l;0h (≈11 p.m. Alexandria), magnitude 6.ld. P.V. Neugebauer gives mid-eclipse as ca. 23.6h Babylon, magnitude 6.1d. The time used by Ptolemy is clearly in error (although the computed positions of sun and moon must have seemed to him to confirm it), but the source of his error is too complicated to discuss here. The best treatment is in Britton[!] 81-4. For this eclipse (alone of those preserved in Almagest) there is also an extant cuneiform report (published by Kugler, SSB I p. 71). According to A.J. Sachs this text should be translated as follows: 'Year VII, month IV, night of the fourteenth, 1⅔ double hours in the night a "total" lunar eclipse took place [with only] a little remaining [uneclipsed]. The north wind blew'. Here the time agrees with modern computations (and disagrees with Ptolemy), but the magnitude disagrees with both.

(for the position of the sun was ♋ 18;12°).

Therefore the lunar position was as follows:

mean position in longitude: ♑ 20;22°

true position in longitude: ♑ 18;14°59

distance [in anomaly] from

the apogee of the epicycle: 28;5°[131]

distance [in latitude] from the northern limit on the inclined circle: 262;12°.

Hence it is clear that, when the centre of the moon, again near its greatest distance, is 7⅘° from the node, as measured along its inclined circle, and the centre of the shadow has the same position relative to it as before, half of the moon's diameter is immersed in the shadow.

But, when the moon's centre is 9⅓° from the node along the inclined circle, it is 48½′ from the ecliptic along the great circle drawn through it at right angles to the inclined circle [the orbit]; and when it is 7⅘° from the node along the inclined circle, it is 4⅔′ from the ecliptic along the great circle drawn through it at right angles to the inclined circle. Therefore, since the difference between [the sizes of] the two eclipses comprises ¼ of the moon's diameter, and the difference between the above distances of the moon's centre from the ecliptic (i.e. from the centre of the shadow) comprises [48½ – 40⅔ =] 7⅚′, it is obvious

131 Ptolemy has made a computing error here: correct is a = 27;54°. Obviously, he has computed (here only) for the uncorrected time of 10⅙h. However, this has no serious consequences, since it is merely intended to show that the moon is near the apogee of the epicycle. The discrepancy in the true position (see n.59) cannot be explained by this error.

that the total diameter of the moon subtends a great circle arc of [4 × 7⅚ =] 31⅓'.

From the same data it is easy to see that the radius of the shadow at the same greatest distance of the moon subtends 40⅔'. For when the moon's centre was that distance [40⅔'] from the centre of the shadow, it was touching the edge of the shadow's circumference, because [in that situation] half of the moon's diameter was eclipsed. This is negligibly Jess than 2⅗ times the radius of the moon, which is 15⅔'. The values we derive for the above quantities from a number of similar observations are in agreement with these;[132] hence we use them, both in other parts of the theory, concerning eclipses,[133] and in the following demonstration of the solar distance, which will be along the same lines as that followed by Hipparchus. A further presupposition [of this demonstration] is that the circles of sun, moon and earth enclosed by the cones are not noticeably less than great circles on their spheres, and the diameters too [not noticeably less than great circle diameters].[134]

---

132 Although Ptolemy's procedure for finding the apparent diameters of moon and shadow is both elegant and theoretically correct, it suffers from serious practical disadvantages. On these, and the inaccuracies involved in his actual computations, see *HAMA* 106–8.

133 Reference to VI 5-7 and VI 11.

134 i.e. in Fig. 5.12 the cones from points N and X enclosing the spheres of sun (ABG), moon (EZH) and earth (KLM) have bases (the circles on AG, EH and KM) which are not sensibly less than great circles in those spheres: thus AG, EH and KM can be treated as diameters of the spheres. This simplifying approximation is fully justified by the magnitude of the distances of the bodies compared with their diameters.

15. On the distance of the sun and other consequences of the demonstration of that[135]

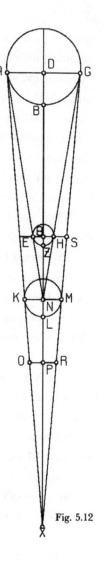

Fig. 5.12

Now, given the above, and given that the greatest distance of the moon at the syzygies is 64;10 units where the earth's radius is 1 (for we showed that its mean distance is 59 of those units, and the radius of the epicyle 5;10), let us see the size of the sun's distance which results.

[See Fig. 5.12.] Let there be the following great circles of the [various] spherical bodies lying in the same plane: circle ABG of the sun's, on centre D, circle EZH of the moon's at its greatest distance, on centre Θ, circle KLM of the earth's, on centre N. Let AXG be the plane through the centres [in the tangent cone] enclosing earth and sun, and ANG the plane through the centres [in the

---

135   On chs. 15 and 16 see *HAMA* 109–12, Pedersen 209–13.

tangent cone] enclosing sun and moon, with DΘNX as common axis. Let the straight lines through the points of tangency, which are, obviously, parallel to each other, and sensibly equal to diameters, be ADG on the sun's circle, EΘH on the moon's circle, KNM on the earth's circle, and OPR on the circle of the shadow in which the moon is immersed at its greatest distance (thus ΘN equals NP, and each of them is 64;10 units where NL, the earth's radius, is 1).

Then we have to find the ratio between ND, the distance of the sun, and NL, the earth's radius.

Produce EH to [meet XG at] S.

Since we demonstrated that the moon's diameter at the distance in question, namely the greatest distance in the syzygies, subtends 0;31,20° of the circle drawn through the moon about the earth's centre,

∠ ENH = 0;31,20° where 4 right angles = 360°,

and ∠ ΘNH = ½ ∠ ENH = 0;31,20°° where 2 right angles = 360°°.

Therefore in the circle about right-angled triangle NHΘ,

arc ΘH = 0;31,20°

and arc ΘN = 1 79;28,40° (supplement).

Therefore the corresponding chords

$$\left. \begin{array}{l} \text{HΘ} = 0;32,48^{\text{P}} \\ \text{and NΘ} \approx 120^{\text{P}} \end{array} \right\} \text{ where diameter NH} = 120^{\text{P}}$$

Therefore where NΘ = 64;10, ΘH = 0;17,33.

And NM, the radius of the earth, is 1 in the same units.

But PR:ΘH ≈ 2;36 : 1.

∴ PR = 0;45,38 in the same units.

∴ ΘH + PR= 1;3,11 where NM = 1.

But PR + ΘS = 2, since PR + ΘS = 2NM

(for, as we said, all [three] are parallel, and NP = NΘ). Therefore, by subtraction [of (PR + ΘH) from (PR + ΘS)],

HS = 0;56,49 where NM = 1.

And NM:HS = NG:HG = ND:ΘD.

Therefore where ND = 1, DΘ = 0;56,49, and, by subtraction, ΘN = 0;3,11.

Therefore where NΘ = 64;10 and NM = 1,

the sun's distance, ND ≈ 1210.

Similarly, as we showed, PR = 0;45,38 where NM = 1, and NM:PR = NX:XP.

Therefore where NX = 1, XP = 0;45,38

and, by subtraction, PN = 0;14,22.

Therefore where PN = 64;10 and NM, the earth's radius, = 1,

XP ≈ 203;50,

and, by addition, XN = 268.

Therefore we have calculated that where the earth's radius is 1 the mean distance of the moon at the syzygies is 59

the distance of the sun is 1210

and the distance from the centre of the earth to the apex of the shadow cone is 268.

16. On the sizes of sun, moon and earth

The ratios of the volumes of the bodies are immediately derivable from the ratios of the diameters of sun, moon and earth.

For, since we have shown that, where NM, the earth's radius, is l,

the moon's radius, ΘH = 0;17,33

and NΘ = 64;10,

and since NΘ:ΘH = ND:DG,

and ND was shown to be 1210 in the same units,

the radius of the sun, DG ≈ 5½ in the same units.

So the diameters will have the same ratios.

Therefore where the moon's diameter is 1, the earth's diameter will be about 3⅖,

and the sun's 18⅘

Therefore the earth's diameter is 3⅖ times the moon's

and the sun's diameter is 18⅘ times the moon's

and 5½ times the earth's.

And, using the same numbers,

since $1^3 = 1$,

and $3⅖^3 ≈ 39¼$,

and $18⅘^3 = 6644½$,

we conclude that, where the moon's volume is 1, the earth's volume is 39¼ and the sun's 6644½.

Therefore the sun's volume is about 170 times that of the earth.[136]

---

136   There is no point in estimating the relative volumes of the bodies, but it was evidently traditional in Greek astronomy, for Théon of Smyrna (ed. Hiller p. 197) and Calcidius (ed. Waszink p. 143) quote from Hipparchus' work on sizes and distances the statement that the sun is 1880 times the size of the earth and the earth 27 times the size of the moon; these ratios plainly refer to relative volumes. In his *Planetary Hypotheses* (ed. Goldstein p. 9) Ptolemy gives the volumes of all the planets relative to the earth.

Book VII

1. That the fixed stars always maintain the same position relative to each other[137]

I n the preceding part of this treatise, Syrus, we discussed the phenomena associated with *sphaera recta* and *sphaera obliqua*, and also the details of the hypotheses for the motions of sun and moon and the combinations of positions which are seen to result from them. Now, to deal with the next part of the theory, we shall begin discussing the stars, and first, in accordance with the logical order, the so-called fixed stars.

First of all we must make the following introductory point. Concerning the terminology we use, in as much as

---

137 On chs. 1 and 2 see Pedersen 237–45.

the stars themselves patently maintain the formations [of their constellations] unchanged and their distances from each other the same, we are quite right to call them 'fixed'; but in as much as their sphere, taken as a whole, to which they are attached, as it were, as they are carried around, also [like the other spheres] has a regular motion of its own towards the rear and the east with respect to the first [daily] motion,[138] it would not be appropriate to call this [sphere] too 'fixed'. For we find that both these statements are true, at least on the [observational] basis afforded by the amount of time [preceding us]: even before this Hipparchus conceived of both these notions on the basis of the phenomena available to him, but under conditions which forced him, as far as concerns the effect over a long period, to conjecture rather than to predict, since he had found very few observations of fixed stars before his own time, in fact practically none besides those recorded by Aristyllos and Timocharis, and even these were neither free from uncertainty nor carefully worked out; but we too come to the same conclusions by comparing present phenomena with those of that time, but with more assurance, both because

---

138 Note that the motion which in modern terminology is 'precession of the equinoxes' (i.e. a motion in the direction of decreasing longitudes of the tropical points with respect to the fixed stars) is described by Ptolemy as a motion of the fixed stars with respect to the tropical points in the direction of increasing longitudes. This accords with his taking the tropical points as the primary reference points (III 1). Hipparchus, however, seems at times to have adopted the modern convention, to judge from the title of his work 'On the displacement of the solsticial and equinoctial points' (III I p. 132 and VII 2 pp. 327 and 329).

our examination is conducted [with material taken] from a longer time-interval, and because the fixed-star observations recorded by Hipparchus, which are our chief source for comparisons, have been handed down to us in a thoroughly satisfactory form.

First, then, no change has taken place in the relative positions of the stars even up to the present time. On the contrary, the configurations observed in Hipparchus' time are seen to be absolutely identical now too. This is true not only of the positions of the stars in the zodiac relative to each other, or of the stars outside the zodiac relative to other stars outside the zodiac (which would [still] be the case if only stars in the vicinity of the zodiac had a rearward motion, as Hipparchus proposes in the first hypothesis he puts forward); but it is also true of the positions of stars in the zodiac relative to those outside it, even those at considerable distances. This can easily be seen by anyone who is willing to make an inspection of the matter and examine, in the spirit of love of truth, whether present phenomena agree with those recorded for Hipparchus' time.

In any case, to provide a convenient test of the matter, we too will adduce here a few of his observations, [namely] those which are most suitable for easy comprehension and also for giving an overview of the whole method of comparison, by showing that the configurations formed by stars outside the zodiac, both with each other and with stars in the zodiac, have been preserved unchanged.[139]

Stars in Cancer. [Hipparchus] records that the star in the southern claw of Cancer [α Cnc], the bright star which is in advance of the latter and of the head of Hydra[β Cnc], and the bright star in Procyon [α CMi] lie almost on a straight line.[140] For the one in the middle lies 1½ digits[141] to the north and east of the[142] straight line joining the two end ones, and the distances [from it to each of them] are equal.

Stars in Leo. [He records] that the easternmost two [μ, ε Leo] of the four stars in the head of Leo [μ, ε, κ, λ], and the star in the place where the neck joins [the head]

---

139   In the following lists I give in brackets the modern designation of the stars in question, when the identification is reasonably certain, and, in footnotes, the equivalent in *Ptolemy's Catalogue of Stars*. Several of the stars mentioned by Hipparchus are not recorded in that catalogue, and his descriptions of those that are often differ from Ptolemy's. In Ptolemy's own alignments which follow, the descriptions also vary somewhat from the catalogue. The alignments are discussed in detail by Manitius, 'Fixsternbeobachtungen'.

140   Catalogue XXV 6 and 9 and XXXIX 2. Like Manitius, I do not understand 'to the north and east'. In the given situation, the only possible deviation is to the north-west or the south-east. I calculate that in Hipparchus' time it was about 5' to the north and west.

141   The 'digit' (δάκτυλος) and 'cubit' (πῆχυς) as astronomical measurements were taken by Hipparchus from Babylonian astronomy (in the Almagest they are found only in the Babylonian observations IX 7, pp. 452–3, and XI 7, p. 541, and in passages derived from Hipparchus). The cubit in Babylonian astronomy can represent either 2½° or 2° (the latter normal in the Hellenistic period: see *HAMA* II 591–93). Strabo, 2.1.18, quotes data from Hipparchus in which the 2° norm is certain. It is also found in Hipparchus' commentary on Aratus, where Vogt, 'Wiederherstellung', col. 30, argued for the 2½° norm. In the passage below, a 2° cubit produces a smaller error in the estimated distance (inaccurate in either case). The 'digit' in Babylonian astronomy is the ¹⁄₂₄th of the 2° cubit or ¹⁄₃₀th of the 2½° cubit, 5' in either case.

142   Reading πῆς for τὴν (misprint in Heiberg) at H4, 14.

of Hydra [ω Hya], lie on a straight line.[143] Also, that the line drawn through the tail of Leo [β] and the star in the end of the tail of Ursa Major [η UMa] cuts off the bright star under the tail of Ursa Major [α CV n] 1 digit to the west [i.e. passes 1 digit to the east of it].[144] Similarly, [he records] that the line through the star under the tail of Ursa Major and the tail of Leo passes through the more advanced of the stars in Coma [Berenices].[145]

Stars in Virgo. [He records] that between the northern foot of Virgo [μ Vir] and the right foot of Bootes [ζ Boo][146] lie two stars; the southern one of these [109 Boo], which is equally bright as the [right] foot of Bootes, lies to the east of the line joining the feet, while the northern one [31 Boo], which is half-bright, lies on a straight line with the feet. Furthermore, of these two stars, the half-bright one is preceded by two bright stars, which form, together with the half-bright one, an isosceles triangle of which the half-bright one is the apex.[147] These [two bright stars] lie on a straight line with Arcturus [α Boo] and the southern foot of Virgo [λ

---

143   Catalogue XXVI 3 and 4 and XLI 6.

144   Catalogue XXVI 27, II 27 and II 28. By my calculation, the line passed more like half a degree to the east of α CVn.

145   The latter are probably catalogue XXVI 33 and 34, doubtfully identified as 15 and 7 Com.

146   Catalogue XXVII 26 and V 19.

147   Manitius identifies these two stars as nos. 43 and 46 of Bootes in the catalogue of Heis (Köln, 1872). I have not tracked these down in a more recent catalogue, since any identification seems utterly uncertain.

Vir].[148] Also, that between Spica [α Vir] and the second star from the end of the tail in Hydra [λ Hya][149] lie three stars, all on one straight line [57,63,69 Vir].[150] The middle one of these [63] lies on a straight line with Spica and the second star from the end of the tail in Hydra.

Stars in Libra. [He records] that the star [μ Ser] which is very nearly on a straight line towards the north with the [two] bright stars in the claws [α, β Lib] is bright and triple: for on both sides of it lie single small stars [36,30 Ser].[151]

Stars in Scorpius. [He records] that the straight line drawn through the rearmost of the stars in the sting of Scorpius [λ Sco] and through the right knee of Ophiuchus [η Oph] bisects the interval between the two advance stars in the right foot of Ophiuchus [36, ϑ Oph][152] and that the fifth and seventh joints [in the tail of Scorpius, ϑ, κ Sco] lie on a straight line with the bright star in the middle of Ara [α Ara].[153] Furthermore, that the northernmost star [δ] of the two in the base of Ara [δ, ϑ][154] lies between and almost on a straight line

---

148 Catalogue V 23 and XXVII 25.

149 Catalogue XXVII 14 and XLI 24.

150 This seems preferable to Manitius' identification (61, 63, 69).

151 The first three are catalogue XIV 11 and XXVIII 1 and 3. My identification of the 'triple star' is far more likely than Manitius' α Ser plus λ, 29 Ser.

152 Catalogue XXIX 20 and XIII 12, 14 and 15.

153 Catalogue XXIX 17 and 19 and XLVI 3.

154 Catalogue XLVI 1 and 2.

with the fifth joint and the star in the middle of Ara, being almost equidistant from both.

Stars in Sagittarius. [He records] that to the east and south of the Circle under Sagittarius [i.e. of Corona Australis] lie two bright stars [α, β Sgr], quite some distance (about 3 cubits) from each other.[155] The southernmost and brighter of these [β], which is on the foot of Sagittarius, lies very nearly on a straight line with the midmost [α CrA] of the three bright stars in the Circle (which lie furthest towards the east in that [constellation]) [γ, α, β CrA], and with the rearmost [ζ Sgr] of the [two] bright stars [ζ, δ Sgr] at opposite angles of the, Quadrilateral [in Sagittarius ζ, τ, δ, φ]: the two intervals [between these three stars] are equal. The northernmost [of the two stars to the east of the Circle, α Sgr] lies to the east of this straight line, but is on a straight line with the [two] bright stars [ζ, δ,] at opposite angles of the Quadrilateral.[156]

4. On the method used to record the positions of the fixed stars

Thus, from our observations and comparisons of the above stars, from similar observations and comparisons of the other bright stars, and from the fact that we found the

---

155 Catalogue XXX 24 and 23. On the cubit see p. 322 n.5.

156 The equivalents in Ptolemy's catalogue are α, β Sgr: XXX 24, 23; γ, α, β CrA: XLVII8, 7,6; ζ, τ, δ, φ Sgr: XXX 22, 21, 6, 7 (*not* described as a quadrilateral).

distances of the other stars with respect to the [bright stars] which we had established to be in agreement [with the results of our predecessors], we have confirmed that the sphere of the fixed stars, too, has a movement towards the rear with respect to the solsticial and equinoctial points of the amount determined (in so far as the time [for which observations are] available allows); furthermore, [we have confirmed] that this motion of theirs takes place about the poles of the ecliptic, and not those of the equator (i.e. the poles of the first motion). So we thought it appropriate, in making our observations and records of each of the above fixed stars, and of the others too, to give their position, as observed in our time, in terms of longitude and latitude, not with respect to the equator, but with respect to the ecliptic, [i.e.] as determined by the great circle drawn through the poles of the ecliptic and each individual star. In this way, in accordance with the hypothesis of their motion established above, their positions in latitude with respect to the ecliptic must necessarily remain the same, while their positions in longitude must always traverse equal arcs towards the rear in equal times.

Hence, again using the same instrument [as we did for the moon, V1], (because the astrolabe rings in it are constructed to rotate about the poles of the ecliptic), we observed as many stars as we could sight down to the sixth magnitude. [We proceeded as follows.] We always arranged the first of the above-mentioned astrolabe rings [Fig. F,5] [to sight] one of the bright stars whose position we had previously determined by means of the moon, setting the ring to the proper graduation on the ecliptic [ring (Fig. F,3) for that star], then set the other ring [Fig. F,2], which was graduated along its entire length and could also be rotated in latitude towards the poles of the ecliptic,[157] to the required star, so that at the same time as the control star was sighted [in its proper position], this star too was sighted through the hole on its own ring. For when these conditions were met, we could readily obtain both coordinates of the required star at the same time by means of its astrolabe ring [Fig. F,2]: the position in longitude was defined by the intersection of that ring and the ecliptic [ring], and the position in latitude by the arc of the astrolabe ring cut off between the same intersection and the upper[158] sighting-hole.

In order to display the arrangement of stars on the solid globe[159] according to the above method, we have set it out below in the form of a table in four sections. For each star (taken by constellation), we give, in the first section, its description as a part of the constellation;[160]

---

157   If the text is sound, Ptolemy is speaking carelessly here. As is clear from the description at V l, ring no. 2 is indeed graduated, but cannot perform a latitudinal movement; that is done by ring no. l, which fits inside no. 2 and has the sighting-holes attached to it.

158   Literally 'above the earth'.

159   For a description of this instrument see VIII 3.

160   Literally 'the shapes' (τάς μορφώσεις), i.e. its position as a part of the mythological figure (animal, anthropomorphic or inanimate) which was delineated on the globe and (notionally) in the heavens.

THE BOOK OF ASTRONOMY IN ANTIQUITY

in the second section, its position in longitude, as derived from observation, for the beginning of the reign of Antoninus[161] ([the position is given] within a sign of the zodiac, the beginning of each quadrant of the zodiac being, as before, established at [one of] the solsticial or equinoctial points); in the third section we give its distance from the ecliptic in latitude, to the north or south as the case may be for the particular star; and in the fourth, the class to which it belongs in magnitude. The latitudinal distances will remain always unchanged, and the positions in longitude can provide a ready means of determining the [corresponding] longitude at other points in time, if we [calculate] the distance in degrees between the epoch and the time in question on the basis of a motion of 1° in 100 years, [and] subtract it from the epoch position for earlier times, but add it to the epoch position[162] for later times.

For the same reasons, our indications [of relative positions] in the descriptions must also be understood to accord with the above kind of hypothesis about the arrangement of the stars, and with the definition [of position] by [circles drawn] through the poles of the ecliptic. Thus, when we speak of a star as 'in advance of' or 'to the rear of' another, we mean that it occupies the relative position in question as defined by the ecliptic

---

161 I.e. according to the Canon Basileon, Thoth 1 of Nabonassar 885 (= 137 July 20).

162 Reading ταίς τῆς επί τοῦ μεταγενεστέρον (with D,Ar) at H37,2 for ταίς τοῦ μεταγενεστέρου. Corrected by Manitius.

position [of the two stars, 'in advance of'] referring to the section of the ecliptic which is in advance, and ['to the rear'] referring to the section of the ecliptic which is towards the rear;[163] and by 'more to the south' or 'more to the north', we mean nearer to the pole of the ecliptic (southern or northern as the case may be). Furthermore, the descriptions which we have applied to the individual stars as parts of the constellation are not in every case the same as those of our predecessors (just as their descriptions differ from their predecessors'): in many cases our descriptions are different because they seemed to be more natural and to give a better proportioned outline to the figures described. Thus, for instance, those stars which Hipparchus places 'on the shoulders of Virgo' we describe as 'on her sides',[164] since their distance from the stars in her head appears greater than their distance from the stars in her hands, and that situation fits [a location] 'on her side', but is totally inappropriate to [a location] 'on her shoulders'. However, one has a ready means of identifying those stars which are described differently [by others]; this can be done immediately simply by comparing the recorded positions.

---

163   Although this is in general true, there appear to be exceptions. See Introduction (on catalogue III 15–18) and n.35 (on catalogue XXXII 23–24).

164   Thus δ Vir is described by Hipparchus (Comm. in Arat. 2.5.5., ed. Manitius p.190,10) as 'the northern shoulder of Virgo', and by Ptolemy (catalogue XXVII 10) as 'the star in the right side under the girdle'.

178

Book IX

1. On the order of the spheres of sun, moon and the 5 planets

Such, then, more or less, is the sum total of the chief topics one may mention as having to do with the fixed stars, in so far as the phenomena [observed] up to now provide the means of progress in our understanding. There remains, to [complete] our treatise, the treatment of the five planets. To avoid repetition we shall, as far as possible, explain the theory of the latter by means of an exposition common [to all five], treating each of the methods [for all planets] together.

First, then, [to discuss] the order of their spheres, which are all situated [with their poles] nearly coinciding with the poles of the inclined, ecliptic circle: we see that almost all the foremost astronomers agree that all the spheres are closer to the earth than that of the fixed stars, and farther from the earth than that of the moon, and that those of the three [outer planets] are farther from the earth than those of the other [two] and the sun, Saturn's being greatest, Jupiter's the next in order towards the earth, and Mars' below that. But concerning the spheres of Venus and Mercury, we see that they are placed below the sun's by the more ancient astronomers, but by some of their successors these too are placed above [the sun's],[165] for the reason that the sun has never been obscured by them [Venus and Mercury] either. To us, however, such a criterion seems to have an element of uncertainty, since it is possible that some planets might indeed be below the sun, but nevertheless not always be in one of the planes through the sun and our viewpoint, but in another [plane], and hence might not be seen passing in front of it, just as in the case of the moon, when it passes below [the sun] at conjunction, no obscuration results in most cases.[166]

---

165 There is a good deal of evidence for the identity of some of those who held the second opinion, including Plato, Eratosthenes and Archimedes. For details on this and other ancient arrangements see *HAMA* II 690–3.

166 I.e. no transits of Venus or Mercury had been observed. Neugebauer has shown (*HAMA* 227–30) that transits are in fact predictable from Ptolemy's own theory. Ptolemy later seems to have realised this, for in the *Planetary Hypotheses* (ed. Goldstein 2,28,10–12) he says: 'if a body of such small size (as a planet) were to occult a body of such large size and with so much light (as the sun), it would necessarily be imperceptible, because of the smallness of the occulting body and the state of the parts of the sun's body which remain uncovered. (Goldstein's translation here, p.6, is inaccurate).

And since there is no other way, either, to make progress in our knowledge of this matter, since none of the stars[167] has a noticeable parallax (which is the only phenomenon from which the distances can be derived), the order assumed by the older [astronomers] appears the more plausible. For, by putting the sun in the middle, it is more in accordance with the nature [of the bodies] in thus separating those which reach all possible distances from the sun and those which do not do so, but always move in its vicinity; provided only that it does not remove the latter close enough to the earth that there can result a parallax of any size.[168]

2. On our purpose in the hypotheses of the planets

So much, then, for the arrangements of the spheres. Now it is our purpose to demonstrate for the five planets, just as we did for the sun and moon, that all their apparent anomalies can be represented by uniform circular motions, since these are proper to the nature

---

167   This includes both fixed stars and planets.

168   In his *Planetary Hypotheses* (see Goldstein's edition) Ptolemy proposes a system in which the spheres of the planets are contiguous; thus the greatest distance from the earth attained by a planet is equal to the least distance attained by the one next in order outwards. This appears to provide support for the order he adopts here, since it results in a solar distance very nearly the same as that obtained by a different method in *Almagest* V 15. Since this system also brings Mercury, at its least distance, to the moon's greatest distance (64 earth-radii), Mercury ought to exhibit a considerable parallax, contrary to what is enunciated here.

of divine beings, while disorder and non-uniformity are alien [to such beings]. Then it is right that we should think success in such a purpose a great thing, and truly the proper end of the mathematical part of theoretical philosophy.[169] But, on many grounds, we must think that it is difficult, and that there is good reason why no-one before us has yet succeeded in it.[170] For, [firstly], in investigations of the periodic motions of a planet, the possible [inaccuracy] resulting from comparison of [two] observations (at each of which the observer may have committed a small observational error) will, when accumulated over a continuous period, produce a noticeable difference [from the true state] sooner when the interval [between the observations] over which the examination is made is shorter, and less soon when it is longer. But we have records of planetary observations only from a time which is recent in comparison with such a vast enterprise: this makes prediction for a time many times greater [than the interval for which observations are available] insecure. [Secondly], in investigation of the anomalies, considerable confusion stems from the fact that it is apparent that each planet exhibits two anomalies, which are moreover

---

169 Cf. I 1.

170 We cannot doubt that not only planetary theories but planetary tables had been constructed before Ptolemy: the proof is supplied by Indian astronomy, which is based on Greek theories which are largely, if not entirely, pre-Ptolemaic, and indeed by Ptolemy's own reference to the 'Aeon tables'. What he means is that all previous efforts were, by his criteria, unsatisfactory.

unequal both in their amount and in the periods of their return: one [return] is observed to be related to the sun, the other to the position in the ecliptic; but both anomalies are continuously combined, whence it is difficult to distinguish the characteristics of each individually. [It is] also [confusing] that most of the ancient [planetary] observations have been recorded in a way which is difficult to evaluate, and crude. For [1] the more continuous series of observations concern stations and phases [i.e. first and last visibilities].[171] But detection of both of these particular phenomena is fraught with uncertainty: stations cannot be fixed at an exact moment, since the local motion of the planet for several days both before and after the actual station is too small to be observable; in the case of the phases, not only do the places [in which the planets are located] immediately become invisible together with the bodies which are undergoing their first or last visibility, but the times too can be in error, both because of atmospherical differences and because of differences in the [sharpness of] vision of the observers. [2] In general, observations [of planets] with respect to one of the fixed stars, when taken over a comparatively great distance, involve difficult computations and an

---

171   Ptolemy is certainly thinking of the Babylonian planetary observations, which are characteristically of this type. They have become available to us through the 'diaries' (see Sachs[2]), but to Ptolemy were probably known only through Hipparchus' compilation (see p. 421).

element of guesswork in the quantity measured, unless one carries them out in a manner which is thoroughly competent and knowledgeable. This is not only because the lines joining the observed stars do not always form right angles with the ecliptic, but may form an angle of any size (hence one may expect considerable error in determining the position in latitude and longitude, due to the varying inclination of the ecliptic [to the horizon frame of reference]); but also because the same interval [between star and planet] appears to the observer as greater near the horizon, and less near mid-heaven;[172] hence, obviously, the interval in question can be measured as at one time greater, at another less than it is in reality.

Hence it was, I think, that Hipparchus, being a great lover of truth, for all the above reasons, and especially because he did not yet have in his possession such a groundwork of resources in the form of accurate observations from earlier times as he himself has provided to us,[173] although he investigated the theories of the sun and moon, and, to the best of his ability,

172 This appears to be the only reference to the effect of refraction (if that is what it is) in the *Almagest*, despite its obvious relevance e.g. to the observations of Mercury's greatest elongations in IX 7. Ptolemy discusses it (as a theoretical problem) in some detail in *Optics* V 23–30 (ed. Lejeune 237–42).

173 This seems to imply that Hipparchus recorded planetary observations of his own, which Ptolemy used to establish his theories. This may be true, but it is strange that Ptolemy cites not a single such observation by Hipparchus. Could Ptolemy mean merely that Hipparchus had not 'yet' assembled the compilation of earlier planetary observations which he mentions just below?

demonstrated with every means at his command that they are represented by uniform circular motions, did not even make a beginning in establishing theories for the five planets, not at least in his writings which have come down to us.[174] All that he did was to make a compilation of the planetary observations arranged in a more useful way,[175] and to show by means of these that the phenomena were not in agreement with the hypotheses of the astronomers of that time. For, we may presume, he thought that one must not only show that each planet has a twofold anomaly, or that each planet has retrograde arcs which are not constant, and are of such and such sizes (whereas the other astronomers had constructed their geometrical proofs on the basis of a single unvarying anomaly and retrograde arc); nor [that it was sufficient to show] that these anomalies can in fact be represented either by means of eccentric circles or by circles concentric with the ecliptic, and carrying epicycles, or even by combining both, the ecliptic anomaly being of such and such a size, and the synodic anomaly of such and such (for these representations have been employed by almost all those who tried to exhibit the uniform circular motion by means of the so-called 'Aeon-

---

174   The circulation of books in antiquity was so fortuitous that, even for one, like Ptolemy, who had access to the great resources of the libraries at Alexandria, this was a necessary caveat.

175   I have little doubt that all the older planetary observations cited in the *Almagest* are derived from this compilation, and that part of Hipparchus' 'rearrangement' was to give their dates in the Egyptian calendar. For a similar service he rendered for the listing of lunar eclipses see *HAMA* 320–21.

tables',[176] but their attempts were faulty and at the same time lacked proofs: some of them did not achieve their object at all, the others only to a limited extent); but, [we may presume], he reckoned that one who has reached such a pitch of accuracy and love of truth throughout the mathematical sciences will not be content to stop at the above point, like the others who did not care [about the imperfections]; rather, that anyone who was to convince himself and his future audience must demonstrate the size and the period of each of the two anomalies by means of well-attested phenomena which everyone agrees on, must then combine both anomalies, and discover the position and order of the circles by which they are brought about, and the type of their motion; and finally must make practically all the phenomena fit the particular character of the arrangement of circles in his hypothesis. And this, I suspect, appeared difficult even to him.

---

176   διὰ τῆς καλουμένης αἰωνίου κανονοποίας. In my opinion, Ptolemy is referring to a type of work in which the mean motions of the planets were represented by integer numbers of revolutions in some huge period, in which they all return to the beginning of the zodiac, and the planetary equations were calculated by a combination of epicycles or of eccentre and epicycle which was not reducible to a geometrically consistent kinematic model, i.e. to a class of Greek works which were the ancestors of the Indian siddhantas. In this I am in agreement with van der Waerden, 'Ewige Tafeln', except that I believe that the αἰών implied by the title of these tables does not mean 'eternity' (cf. the conventional translation, 'Eternal Tables', which is philologically possible, but not necessary), but refers to the immense common period in which the planets return (cf. the Greek inscription of Keskinto, *HAMA* 698–705, and the Indian Mahayuga). The other two references to these tables in antiquity (P. Lond. 130, see Neugebauer-van Hoesen, *Greek Horoscopes* p. 21, I 12–13, and Vettius Valens VI I, ed. Kroll 243,8) are consistent with, but do not require, this interpretation.

The point of the above remarks was not to boast [of our own achievement]. Rather, if we are at any point compelled by the nature of our subject to use a procedure not in strict accordance with theory (for instance, when we carry out proofs using without further qualification the circles[177] described in the planetary spheres by the movement [of the body, i.e.] assuming that these circles lie in the plane of the ecliptic,[178] to simplify the course of the proof); or [if we are compelled] to make some basic assumptions which we arrived at not from some readily apparent principle, but from a long period of trial and application,[179] or to assume a type of motion or inclination of the circles which is not the same and unchanged for all planets;[180] we may [be allowed to] accede [to this compulsion], since we know that this kind of inexact procedure will not affect the end desired, provided that it is not going to result in any noticeable error; and we know too that assumptions made without proof, provided only that they are found to be in agreement with the phenomena, could not have been found without some careful methodological procedure, even if it is difficult to explain how one came to conceive them (for, in general, the cause of first principles is, by nature, either

---

177 Literally 'as if the circles were bare [circles]'.

178 Ptolemy in fact carries out all the proofs involving the longitudinal motions of the planets (in Bks. IX-XII) as if the motions lay in the plane of the ecliptic.

179 The paradigm case of this is the introduction of the equant.

180 E.g. the special model for the longitudinal motions of Mercury, or the special inclinations attributed to the inner planets for their latitudinal motions.

non-existent or hard to describe); we know, finally, that some variety in the type of hypotheses associated with the circles [of the planets] cannot plausibly be considered strange or contrary to reason (especially since the phenomena exhibited by the actual planets are not alike [for all]); for, when uniform circular motion is preserved for all without exception, the individual phenomena are demonstrated in accordance with a principle which is more basic and more generally applicable than that of similarity of the hypotheses [for all planets].

The observations which we use for the various demonstrations are those which are most likely to be reliable, namely [1] those in which there is observed actual contact or very close approach to a star or the moon, and especially [2] those made by means of the astrolabe instruments. [In these] the observer's line of vision is directed, as it were, by means of the sighting-holes on opposite sides of the rings, thus observing equal distances as equal arcs in all directions, and can accurately determine the position of the planet in question in latitude and longitude with respect to the ecliptic, by moving the ecliptic ring on the astrolabe, and the diametrically opposite sighting-holes on the rings[181] through the poles of the ecliptic, into alignment with the object observed.

---

181   It is not clear why the plural ('rings') is used (contrast the singular at V1, H354,13). Although the sights are attached only to ring 1 in Fig. F. Ptolemy is presumably referring to both ring 1 and ring 2, since ring 2 has first to be moved to the correct sighting position on the ecliptic ring (no. 3).

5. Preliminary notions [necessary] for the hypotheses of the 5 planets[182]

Now that these [mean motions] have been tabulated, our next task is to discuss the anomalies which occur in connection with the longitudinal positions of the five planets. The way we have approached it, to give the general outlines, is as follows.

There are, as we said,[183] two types of motion which are simplest and at the same time sufficient for our purpose; [namely] that produced by circles eccentric to [the centre of] the ecliptic, and that produced by circles concentric with the ecliptic but carrying epicycles around. There are likewise two apparent anomalies for each planet: [1] that anomaly which varies according to its position in the ecliptic, and [2] that which varies according to its position relative to the sun.

For [2] we find, from a series of different [sun-planet] configurations observed round about the same part of the ecliptic,[184] that in the case of the five planets[185] the

---

182 On chs. 5 and 6 see *HAMA* 149–50.

183 III 3.

184 This eliminates the effect of the ecliptic anomaly. Examples would be observations of Mars at opposition, station and (by interpolation) conjunction all near the same point in the ecliptic.

185 Excising καί before ἐπί τῶν πέντε πλανωμένων at H250,17. (καί was apparently omitted in the text translated by al-Hahhaj). One would have to translate Heiberg's text 'in the case of the five planets too' (as well as the sun and moon). But the situation is precisely the opposite for the sun and moon (see e.g. III 4). Perhaps the whole phrase καί ... πλανωμένων is an ancient interpolation.

time from greatest speed to mean is always greater than the time from mean speed to least. Now this feature cannot be a consequence of the eccentric hypothesis, in which exactly the opposite occurs, since the greatest speed takes place at the perigee in the eccentric hypothesis, while the arc from the perigee to the point of mean speed is less than the arc from the latter to the apogee in both [eccentric and epicyclic] hypotheses. But it can occur as a consequence of the epicyclic hypothesis, however only when the greatest speed occurs, not at the perigee, as in the case of the moon, but at the apogee; that is to say, when the planet, starting from the apogee, moves, not as the moon does, in advance [with respect to the motion] of the universe, but instead towards the rear. Hence we use the epicyclic hypothesis to represent this kind of anomaly.[186]

But for [1], the anomaly which varies according to the position in the ecliptic, we find from [observations of] the arcs of the ecliptic between [successive] phases or [sun-planet] configurations of the same kind[187] that the opposite is true: the time from least speed to mean is always greater than the time from mean speed to greatest. This feature can indeed be a consequence of either of the two

---

186 See Ptolemy's discussion of this point at III 3. However, as Neugebauer points out (*HAMA* 149–50) it is perfectly possible for an eccentric model to represent the planetary motions, provided the apsidal line is allowed to move, and precisely this kind of eccentric model is described at XII I, though even there Ptolemy restricts its applicability to the outer planets.

187 This eliminates the effect of the synodic anomaly. Examples would be observations of oppositions of Mars in different parts of the ecliptic (as in X 7).

hypotheses (in the way we described in our discussion of the equivalence of the hypotheses at the beginning of our treatise on the sun [III 3]). But it is more appropriate to the eccentric hypothesis,[188] and that is the hypothesis which we actually use to represent this kind of anomaly, since, moreover, the other anomaly was found to be peculiar, so to speak, to the epicyclic hypothesis.

Now from prolonged application and comparison of observations of individual [planetary] positions with the results computed from the combination of both [the above] hypotheses, we find that it will not work simply to assume[189] [as one has hitherto] that the plane in which we draw the eccentric circles is stationary, and that the straight line through both centres (the centre of the [planet's] eccentre and the centre of the ecliptic), which defines apogee and perigee, remains at a constant distance from the solstitial and equinoctial points; nor [to assume] that the eccentre on which the epicycle centre is carried is identical with the eccentre with respect to the centre of which the epicycle makes its uniform revolution towards the rear, cutting off equal angles in equal times at [that centre]. Rather, we find that the apogee of the eccentre performs a slow motion towards the rear with respect to the solstices, which is uniform about the centre of the ecliptic, and comes to about the same for each planet as the amount determined for the

---

188   Cf. III 4, where Ptolemy prefers it on the ground that it is 'simpler'.

189   Literally 'that the assumption that ... cannot progress so simply'.

sphere of the fixed stars, i.e. 1° in 100 years (at least, as far as can be estimated on the basis of available evidence). We find, too, that the epicycle centre is carried on an eccentre which, though equal in size to the eccentre which produces the anomaly, is not described about the same centre as the latter. For all planets except Mercury the centre [of the actual deferent] is the point bisecting the line joining the centre of the eccentre producing the anomaly to the centre of the ecliptic. For Mercury alone, [the centre of the deferent] is a point whose distance from the centre of the circle about which it rotates is equal to the distance of the latter point towards the apogee from the centre of the eccentre producing the anomaly, which in turn is the same distance towards the apogee from the point representing the observer; for also, in the case of this planet alone, we find that, just as for the moon, the eccentre is rotated by [the movement of] the above-mentioned centre in the opposite sense to the epicycle, [i.e.] in the advance direction, one rotation per year. [This must be so] because the planet itself appears twice in the perigee in the course of one revolution, just as the moon appears twice in the perigee in one [synodic] month.

6. On the type of and difference between the hypotheses

One may more easily grasp the type of the hypotheses which we infer on the basis of the preceding [phenomena] from the following description.

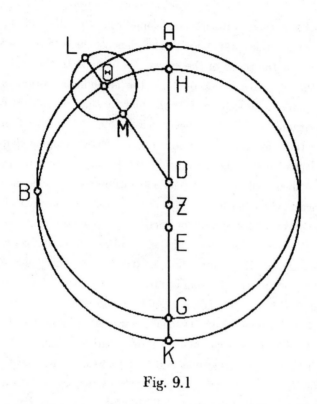

Fig. 9.1

First for that of the [four planets] other [than Mercury],
imagine [Fig. 9.1] the eccentre ABG about centre D, with ADG
as the diameter through D and the centre of the ecliptic; on this
let E be taken as the centre of the ecliptic, i.e. the viewpoint of
the observer, making A the apogee and G the perigee. Let DE
be bisected at Z and with centre Z and radius DA draw a circle
HΘK, which must, clearly, be equal to ABC. Then on centre Θ
draw the epicycle LM, and join LΘMD.

First, then, although we assume that the plane of the eccentric circles is inclined to the plane of the ecliptic, and also that the plane of the epicycle is inclined to the plane of the eccentres, to account for the latitudinal motion of the planets, in accordance with what we shall demonstrate concerning that topic, nevertheless, for the motions in longitude, for the sake of convenience, let us imagine that all [those planes] lie in a single [plane], that of the ecliptic, since there will be no noticeable longitudinal difference, not at least when the inclinations are as small as those which will be brought to light for each of the planets. Next, we say that the whole plane [of the eccentre] moves uniformly about centre E towards the rear [i.e. in the order] of the signs, shifting the position of apogee and perigee 1° in 100 years, and that diameter LEΘM of the epicycle rotates uniformly about centre D, again towards the rear [i.e. in the order] of the signs, with a speed corresponding to the planet's return in longitude, and that it carries with it points L and M of the epicycle, and centre Θ of the epicycle (which always moves on the eccentre HEΘK), and also carries with it the planet; the planet, for its part, moves with uniform motion on the epicycle LM and performs its return always with respect to that diameter [of the epicycle] which points towards centre D, with a speed corresponding to the mean period of the synodic anomaly, and [a sense of rotation] such that its motion at the apogee L takes place towards the rear.

194

Book X

1. Demonstration of [the position of] the apogee of the planet Venus[190]

Such, then, was the method by which we found the hypotheses for the planet Mercury, the sizes of its anomalies, and also the precise amounts of its periodic motions, and their epochs. For the planet Venus, again, we first investigated the position in the ecliptic of the apogee and perigee of the eccentre by [finding] greatest elongations which are equal and in the same direction.[191] The available ancient observations did

---

190 On chapters 1–3 see *HAMA* 152–6. Pedersen 298–306 and (for a criticism of Ptolemy's procedure) Sawyer, 'Ptolemy's determination of the apsidal line for Venus' (cf. p. 449 n.53).

191 Many of the dates of greatest elongations of Venus given here by Ptolemy are in error, some by as much as three weeks (see *HAMA* 153 n.l). We cannot doubt that he was aware of this, but he was forced by the lack of suitable observations during the limited period available to take those positions of Venus close to greatest elongation which gave the required positions of the mean sun with respect to Venus' apsidal line.
The point is discussed in detail by Swerdlow and Neugebauer, Ch.5.

not supply us with exact pairs of positions [suitable] for this purpose, but we used contemporary observations for our approach, as follows.

1. Among the observations given to us by the mathematician Théon, we found one recorded in the sixteenth year of Hadrian, on Pharmouthi [VIII] 21/22 in the Egyptian calendar [132 Mar. 8/9), at which, he says, the planet Venus was at its greatest elongation as evening-star from the sun, and was the length of the Pleiades in advance of the middle of the Pleiades; and it seemed to be passing it a little to the south. Now, according to our coordinates, the longitude of the middle of the Pleiades at that time was ♉ 3°, and its length is about 1½°:[192] so clearly Venus' longitude at that moment was ♉ 1½°. So, since the longitude of the mean sun at that moment was ♓ 14¼°, the greatest distance from the mean as evening-star was 47¼°.

2. In the fourth[193] year of Antoninus, Thoth [I] 11/12 in the Egyptian calendar [140 July 29/30], we observed Venus at its greatest elongation from the sun as morning-star. It was [the breadth of] half a full moon to the north-east of [the star

---

192 In the catalogue (XXIII 30–32) the group of the Pleiades has longitudes between ♉ 2 ⅙° and ♉ ⅔°. The length of this is indeed 1½°, but its midpoint is ♉ 2;55°, which Ptolemy has rounded to 3° (a correction lor precession would make it even less than 2;55°).

193 Reading ιθ′ (with D,Ar) for ιη′ (14th) at H297,5. The date is confirmed by the computations below. Corrected by Manitius.

THE BOOK OF ASTRONOMY IN ANTIQUITY

in] the middle knee of Gemini. At that moment the longitude of the fixed star, according to us, was Ⅱ 18¼°,[194] so Venus was in about Ⅱ 18½°. And the mean sun was in ♌ 5¾°. So the greatest distance as morning-star was the same amount as before, 47¼°. Therefore, since the mean position was ♓ 14¾° at the first observation, and ♌ 5¾° at the second, and the point on the ecliptic halfway between these falls in [either] ♉ 25° [or] ♏ 25°, the diameter through apogee and perigee must go through the latter [points].

3. Similarly, in the [observations we got] from Théon, we found that in the twelfth year of Hadrian, Athyr [III] 21/22 in the Egyptian calendar (127 Oct. 11/12), Venus as morning-star had its greatest elongation from the sun when it was to the rear of the star on the tip of the southern wing of Virgo by the length of the Pleiades, or less than that amount by its own diameter; and it seemed to be passing the star one moon to the north. Now the longitude of the fixed star at that time, according to us, was ♌ 28¹¹⁄₁₂°: hence the longitude of Venus was about ♍ 0⅓°.[195] And the mean sun was ♎

---

194 Catalogue XXIV 11, where the description is somewhat different. Of the three knees mentioned (nos. 10, 11 and 13) this is the middle one.

195 Literally 'a third of the first degree of Virgo'. The longitude in the catalogue (XXVII 5) is ♌ 29°. Ptolemy subtracts 5' for 11 years' precession, adds 1½° for the length of the Pleiades, and then subtracts 5' for the diameter of Venus. (In the Planetary Hypotheses, ed. Goldstein p. 8 § 5, he estimates the apparent diameter of Venus as ¹⁄₁₀th of the sun's, i.e. 3').

197

17²⁶/₃₀°. So the greatest elongation from the mean as morning-star was 47¹⁶/₃₀°.

4. In the twenty-first year of Hadrian, Mechir [VI] 9/10 in the Egyptian calendar [136 Dec. 25/26], in the evening, we observed Venus at its greatest elongation from the sun. It was in advance of the northernmost star of the four which almost form a quadrilateral (behind the star to the rear of and on a straight line with the [two] in the groin of Aquarius):¹⁹⁶ [its distance from the star was] about two-thirds of a full moon, and it seemed about to obscure the star with its light.¹⁹⁷ Now the longitude of the fixed star at that time, according to us, was ♒ 20°; hence Venus was in about ♒ 19³/₅°,¹⁹⁸ and the mean sun's longitude was ♑ 2¹/₁₅°. Here too, then, the greatest elongation as evening-star was the same [as in [3] as morning-star], 47¹⁶/₃₀°. And the points on the ecliptic halfway between the ♎ 17²⁶/₃₀° of the first observation and the ♑ 2¹/₁₅° of the second are again about ♏, 25° and ♉ 25°.

---

196 The stars in question are (according to Manitius' identification): the quadrilateral, catalogue nos. XXXII 26–9; the two in the groin, nos. 15 and 16. The differences in the description here from the catalogue are so great that we must assume that this was originally written before the catalogue existed (as the date of the observation suggests).

197 Reading καταλάμψειν (with GD) for καταλάμπειν ('seemed to be obscuring') at H298,14–15. The word is a technical term for one bright body (the sun, as at VIII 6, H201, 1, cf. καταλάμψεις XIII 7, H591, 11, or the moon, as here) coming so close to another that it 'outshines' it and makes it no longer visible.

198 'two-thirds of a moon' is only 20', whereas Ptolemy subtracts 24'. Is the difference to account for the diameter of Venus?

2. On the size of [Venus'] epicycle

By these means, then, we determined that in our time the apogee and perigee of [Venus'] eccentre lie in ♉ 25° and ♏, 25°. Accordingly, we again looked for greatest elongations from the mean which occur when the sun is near ♉ 25° and ♏ 25°.

1. In the [observations] given to us by Théon we find that in the thirteenth year of Hadrian, Epiphi [XI] 2/3 in the Egyptian calendar (129 May 19/20), Venus was at its greatest elongation from the sun as morning-star, and was 1²/₅° in advance of the straight line through the foremost of the 3 stars in the head of Aries and the star on the hind leg, while its distance from the foremost star of those in the head was approximately double its distance from the star on the leg. Now at that time, according to us, the foremost star of the 3 in the head of Aries had a longitude of [♈] 6³/₅° and is 7¹/₃° north of the ecliptic, while the star in the hind leg of Aries had a longitude of 14¾°, and is 5¼° south of the ecliptic.[199] Therefore the

---

199   The stars in question are catalogue XXII I and 13 (note the different descriptions there), with longitudes of 6⅔° and 15°. The difference in the longitudes given here is -4' and -15' respectively. One would expect about -5' for the precession in 8 years. Hence Manitius emended 14¾ to 14¹¹/₁₂; but it is implausible to change, as he does. ∠' δ ' to 3 ' (⅔ + ¼); for ¹¹/₁₂ is written ∠ ' γ ' ιβ ' (½ + ⅓ + ¹/₁₂), e.g. H303,7. The stars in the alignment are too far apart to allow us to use it to check the text, so in the absence of any ms. variation I merely note the possibility of some corruption.

longitude of Venus was ♈ 10⅗° and it was 1½° south of the ecliptic. Hence, since the longitude of the mean sun at that time was ♉ 25⅖°, the greatest elongation from the mean was 44⅘°.

2. In the twenty-first year of Hadrian, Tybi [V] 2/3 in the Egyptian calendar [136 Nov. 18/19], in the evening, we observed Venus at its greatest distance from the sun: when sighted with respect to the stars in the horns of Capricorn it was seen to occupy ♑ 12⅚°, while the longitude of the mean sun was ♏ 25½°. Hence in this position the greatest elongation from the mean comes out as 47⅓°.

Hence it is clear that the apogee lies in ♉ 25°, and the perigee in ♏ 25°. Furthermore, it has also become plain to us that the eccentre of Venus carrying the epicycle is fixed, since nowhere on the ecliptic do we find the sum of the greatest elongations from the mean on both sides to be less than the sum of both in Taurus, or greater than the sum of both in Scorpius.

With the above as data, let [Fig. 10.1] the eccentric circle, on which Venus' epicycle is always carried, be ABG on diameter AG, on which D is taken as the centre of the eccentre, E as the centre of the ecliptic, and A as the point at ♉ 25°. About points A and G let there be drawn equal epicycles, on which lie points Z and H [respectively]. Draw the tangents EZ and EH, and join AZ, GH.

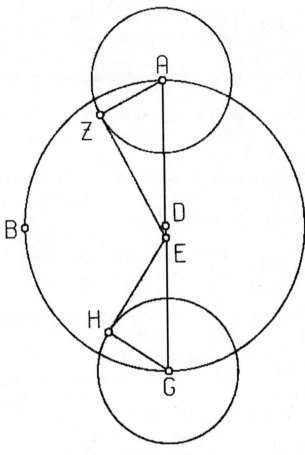

Fig. 10.1

Then, since ∠ AEZ, which is at the centre of the ecliptic, subtends the greatest elongation of the planet at the apogee, which is, by hypothesis, 44⅘°,

$$\angle \text{ AEZ} = \begin{cases} 44;48° \text{ where 4 right angles} = 360° \\ 89;36°° \text{ where 2 right angles} = 360°° \end{cases}$$

Therefore in the circle about right-angled triangle AEZ

arc AZ = 89;36°

and its chord AZ $\approx$ 84;33^P where hypotenuse AE = 120^P.

Similarly, since $\angle$ GEH subtends the greatest elongation at the perigee,

which is, by hypothesis, 47⅓°,

$$\angle \text{ GEH} = \begin{cases} 47;20° \text{ where 4 right angles} = 360° \\ 94;40°° \text{ where 2 right angles} = 360°° \end{cases}$$

Therefore in the circle about right-angled triangle GEH

arc GH = 94;40°

and its chord GH $\approx$ 88;13^P where hypotenuse EG = 120^P.

Therefore where GH (= AZ), the radius of the epicycle, is 84;33^P and AE = 120^P,

EG = 115;1^P,

And obviously, by addition, AG = 235;1^P

and its half, AD $\approx$ 117;30^P,

and, by subtraction, the distance between the centres, DE = 2;29^P. Therefore where the radius of the eccentre, AD = 60^P,

the distance between the centres, DE $\approx$ 1¼^P, and the radius of the epicycle, AZ = 43⅙^P.

3. On the ratios of the eccentricities of the planet [Venus]

But since it is not clear whether the uniform motion of the epicycle takes place about point D, here too we took two greatest elongations, in opposite directions [i.e. one as evening-star and the other as morning-star], in each of which[200] the mean motion of the sun was a quadrant from the apogee.

1. We observed the first in the eighteenth year of Hadrian, Pharmouthi [VIII] 2/3 in the Egyptian calendar (134 Feb. 17/18). In this Venus was at greatest elongation from the sun as morning-star, and when it was sighted with respect to the star called Antares [catalogue XXIX 8], its longitude was ♑, 11¹¹/₁₂°, at which time the longitude of the mean sun was ♒ 25½°. So the greatest elongation from the mean as morning-star was 4⁷/₁₂°.

2. We observed the second in the third year of Antoninus, Pharmouthi [VIII] 4/5 in the Egyptian calendar [140 Feb. 18/19], in the evening. In this Venus was at its greatest elongation from the sun, and when it was sighted with respect to the bright star in the Hyades [catalogue XXIII 14), its longitude was ♈ 13⁵/₆°, while the longitude of the mean sun was again ♒ 25½°. Hence in this case

---

200   Reading ἐφ' ἑκατέρας (with CDG,ls) at H303,2 for ἐφ' ἑκατέρα ('in both directions'). Corrected by Manitius.

the greatest elongation from the mean as evening star was 48⅓°.

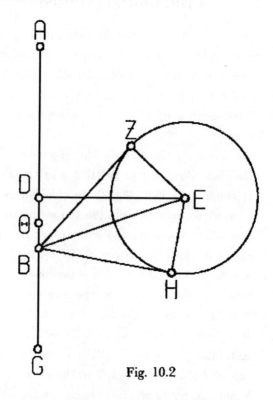

Fig. 10.2

With the above as data, let [Fig. 10.2] the diameter through the apogee and perigee of the eccentre be ABG; let A represent the point at ♉ 25°, and let B represent the centre of the ecliptic. Let our task be to find the centre about which we say that the uniform motion of the epicycle takes place. Let that centre be point D, and

204

draw DE through D perpendicular to AG, in order for the mean position of the epicycle to be a quadrant from the apogee, as in the observations. On DE take E to represent the centre of the epicycle at the observations in question, draw the epicycle ZH on it as centre, draw the tangents to it from B, BZ and BH, and join BE, EZ and EH.

Then since, at the mean position in question, the greatest elongation from the mean as morning-star is, by hypothesis, 43$\frac{7}{12}$°, and the greatest as evening-star 48$\frac{1}{3}$°,

by addition, $\angle$ ZBH = 91;55° where 4 right angles = 360°. Therefore its half, $\angle$ ZBE = 91;55°° where 2 right angles = 360°°.

Therefore in the circle about right-angled triangle BEZ
$$\text{arc EZ} = 91;55°$$
and EZ = 86;16^P where hypotenuse BE = 120^P.

Therefore where the radius of the epicycle, EZ = 43; 10^P
$$\text{BE} = 60;3^P.$$

Again, since the difference between the above greatest elongations (which is 4;45°) comprises twice the equation of the ecliptic anomaly at that point, which is represented by $\angle$ BED,

$$\angle \text{ BED} = \begin{cases} 2;22,30° \text{ where 4 right angles} = 360° \\ 4;45°° \text{ where 2 right angles} = 360°° \end{cases}$$

Therefore in the circle about right-angled triangle BDE
$$\text{arc BD} = 4;45°$$

205

and BD $\approx$ 4;59″ where hypotenuse BE = 120^P. Therefore where BE = 60:3^P and the radius of the epicycle is 43;10^P,

$$BD \approx 2\frac{1}{2}^P.$$

But we showed that the distance between B, the centre of the ecliptic, and the centre of the eccentre on which the epicycle centre is always carried, is 1¼^P in the same units; thus it is half of BD.

Therefore, if we bisect BD at Θ, we have demonstrated[201] that where ΘA, the radius of the eccentre carrying the epicycle, is 60^P, each of the distances between the centres, BΘ and ΘD = 1¾^P,

and EZ, the radius of the epicycle, is 43;10^P.

Q.E.D.

6. Preliminaries for the demonstrations concerning the other [3 outer] planets

Such, then, were the methods which we successfully used for these two planets, Mercury and Venus, to establish the hypotheses and demonstrate [the sizes of] the anomalies. For the other three, Mars, Jupiter and Saturn, the hypothesis which we find for their motion is the same [for

---

201 This is the only 'demonstration' of the 'bisection of the eccentricity' in the *Almagest*, although it is also assumed for the outer planets. However, this does not prove (*contra HAMA* 155) that observations of Venus were the historical origin of Ptolemy's introduction of the equant. It seems far more likely that it arose from the considerations Ptolemy himself outlines at X 6 for which Mars must have provided the most opportune observations.

all] and like that established for the planet Venus, namely
one in which the eccentre on which the epicycle centre is
always carried is described on a centre which is the point
bisecting the line joining the centre of the ecliptic and the
point about which the epicycle has its uniform motion;
for in the case of each of these planets too, using rough
estimation, the eccentricity one finds from the greatest
equation of ecliptic anomaly turns out to be about
twice that derived from the size of the retrograde arcs
at greatest and least distances of the epicycle. However,
the demonstrations by which we calculate the amounts of
both anomalies and [the positions of] the apogees cannot
proceed along the same lines for these planets as for the
previous two, since these reach every possible elongation
from the sun, and it is not obvious from observation, as it
was from the greatest elongations for Mercury and Venus,
when the planet is at the point where the line of our sight
is tangent to the epicycle. So, since that approach is not
available, we have used observations of their oppositions
to the mean position of the sun to demonstrate, first of
all, the ratios of their eccentricities and [the positions of]
their apogees. For only in such positions [of the planet],[202]
considered from a theoretical point of view, do we find the
ecliptic anomaly isolated, with no effect from the anomaly
related to the sun.

---

202   See *HAMA* 172. An ingenious analysis of the way in which Ptolemy arrived at the
notion of the equant for the outer planets was made by Swerdlow, 'The Origin of Ptolemaic
Planetary Theory'.

For let [Fig. 10.5] the planet's eccentre, on which the epicycle centre is carried, be ABG on centre D, and let the diameter through the apogee be AG, on which point E is the centre of the ecliptic, and Z the centre of that eccentre with respect to which the epicycle's mean motion in longitude is taken. Draw the epicycle HΘKL on centre B, and join ZLBΘ and HBKEM.

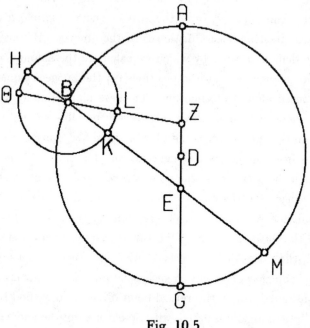

Fig. 10.5

I say, first, that when the planet is seen along line EH through the epicycle centre B, then the mean position of the sun, too, will always be on the same line, and that when the planet is at H it will be in conjunction[203] with the mean sun (which will also, in theory, be seen towards H), and when the planet is at K it will be in opposition to the mean sun (which will be seen, in theory, towards

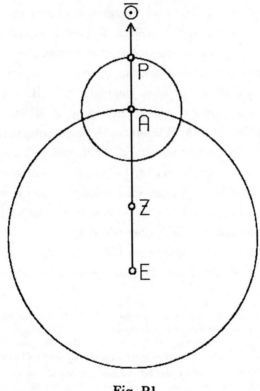

Fig. P1

---

203   Reading συνοδεύσει (with G, and possibly Ar, but the translations are ambiguous) for συνοδεύσει ('is in conjunction') at H318,18.

M). [Proof:] For each of these [outer] planets, the sum of the mean motions in longitude and anomaly, counted from the apogee [of eccentre and epicycle respectively], equals the mean motion of the sun counted from the same starting-point. And the difference between the angle at centre Z (which comprises the mean motion of the planet in longitude), and the angle at E (which comprises the apparent motion in longitude),[204] is always the angle at B (which comprises the mean motion on the epicycle). Hence it is clear that when the planet is at H, it will fall short of a return to the apogee Θ by ∠ HBΘ; but ∠ HBΘ added to ∠ AZB produces the angle comprising the sun's mean motion, namely ∠ AEH, which is the same as the apparent motion of the planet.[205] And when the planet is at K, its motion on the epicycle, again, will be ∠ ΘBK, and ∠ ΘBK + ∠ AZB equal the mean motion of the sun counted from the apogee A.

Thus the latter comprises 180° + (∠ AZB − ∠ LBK) = 180° + ∠ GEM, i.e. the mean position of the sun will

---

204 By this expression (ἡ φαινομένη κατά μήκος κίνηοις) Ptolemy means, not the true position of the planet, but the position of the epicycle center as seen from the earth. Compare the expression ἡ φαινομένη ἐπί τοῦ ἐπικύκλου πάροδος at XII 2 (H470,1 l) to denote the 'true anomaly' (i.e. as counted from true and not mean perigee of the epicycle).

205 In fact ∠ AZB − ∠ HBΘ = ∠ AEH. But what Ptolemy means is illustrated by Figs. P1 and P2: in Fig. P1 planet and mean sun are in conjunction. In Fig. P2 (= Fig. 10.5) they are again in conjunction. The epicycle has travelled through the angle κ̄ (∠ AZB), the planet on the epicycle has travelled through ᾱ and the mean sun through κ̄ + 360°. Then (from the figure) κ = κ̄ (360° − ᾱ) = κ̄ + ᾱ − 360°. Hence the mean sun's motion κ + 360° = κ̄ + ᾱ. Failing to understand this, an interpolator has inserted τουτέστιν λειφθεῖσα ὑπ' αὐτῆς at 319,8, producing the strange result '∠ HBΘ added to ∠ AZB. i.e. subtracted from it.'

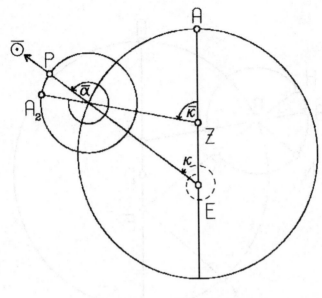

Fig. P2

be opposite the apparent position of the planet. Hence, furthermore, in such configurations [i.e. mean conjunctions and oppositions], the line joining the epicycle centre B to the planet, and the line from E, our point of view, to the mean sun, will coincide in one straight line, but at all other [sun-planet] elongations [those vectors] will always be parallel to each other, although the direction in which they point will vary.

[Proof:] In the below figure [see Fig. 10.6], if we draw the line BN from B to the planet in any situation, and the line EX from E to the mean sun, for the reasons stated below:

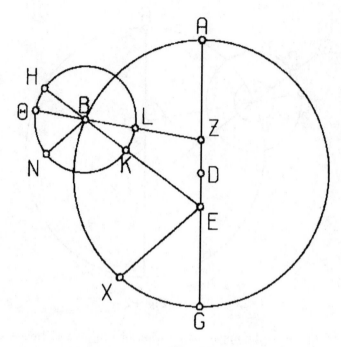

Fig. 10.6

$$\angle \text{ AEX} = \angle \text{ AZ}\Theta + \angle \text{ NB}\Theta,^{206}$$
and $\angle$ AZΘ = $\angle$ AEH + $\angle$ HBΘ.
$$[\therefore \angle \text{ AEX} = \angle \text{ AEH} + \angle \text{ NB}\Theta + \angle \text{ HB}\Theta.]$$
If we subtract $\angle$ AEH from both sides,
$$\angle \text{ HEX} = \angle \text{ HBN}.$$
Therefore line EX is parallel to line BN.

---

206 I.e. the mean motion of the sun equals the mean longitudinal motion of the planet plus the mean anomaly of the planet.

Thus we find that in the above configurations of conjunction and opposition with respect to the mean sun, the planet is viewed, in theory, [along the line] through the centre of the epicycle, just as if its motion on the epicycle did not exist, but instead it were itself situated on circle ABG and were carried in uniform motion by the line ZB, in the same way as the epicycle centre is. Hence it is clear that it is possible to isolate and demonstrate the ratio of the ecliptic eccentricity by [both] such types of [planetary] positions, but since the conjunctions are not visible, we are left with the oppositions[207] on which to build our demonstrations.

7. Demonstration of the eccentricity and apogee [position] of Mars[208]

In the case of the moon we took the positions and times of three lunar eclipses, and demonstrated the ratio of the anomaly and the position of the apogee geometrically. So too, here, in the same way, for each of these [outer] planets, we observed the positions of three oppositions to the mean sun, as accurately as possible, using the astrolabe instruments, computed, too, the time and position for the precise 180° elongation from the position of the mean sun at [each of] the observations, and thence demonstrate the ratio of the eccentricity and [the position of] the apogee.

---

207 ἀκρώνυκτοι σχηματισμοί literally 'configurations [at which the planet rises and sets] at the beginning and end of night'.

208 On the method used to find the eccentricities of the outer planets see *HAMA* 172–7, Pedersen 273–83.

First, then, for Mars, we took three oppositions, which we observed as follows.[209]

1. The first in the fifteenth year of Hadrian, Tybi [V] 26/27 in the Egyptian calendar [130 Dec. 14/15], 1 equinoctial hour after midnight, at about Ⅱ 21°.

2. The second in the nineteenth year of Hadrian, Pharmouthi [VIII] 6/7 in the Egyptian calendar [135 Feb. 21/22], 3 hours before midnight, at about ♌ 28;50°.

3. The third in the second year of Antoninus, Epiphi [XI] 12/13 in the Egyptian calendar [139 May 27/28], 2 equinoctial hours before midnight, at about ♐ 2;34°.

The intervals between the above are as follows:

From oppositions [1] to [2] 4 Egyptian years 69 days 20 equinoctial hours.

From [2] to [3] 4 years 96 days 1 equinoctial hour.

For the first interval we compute a [mean] motion in longitude, beyond complete revolutions, of 81;44°

and for the second interval, 95;28°.

Even if we used the crude periods of return, which we listed above, to compute the mean motions, it

_____

209 The times are arrived at by computing the position of the mean sun. Therefore the computed position of the mean sun at the time stated ought to be exactly 180° different from the longitudes given. I find, from the solar mean motion tables, 260;58,55° (instead of 261°), 328;50,22° (for 328;50°) and 62;31,45° (for 62;34°). The latter discrepancy represents about half an hour in solar motion. Could Ptolemy have applied the equation of time (which is about -25½ mins. compared with epoch) here? If so, he was mistaken, since all the computations are in terms of mean solar days.

would make no significant difference over such a short interval.[210]

It is obvious that the apparent motion of the planet, beyond complete revolutions, is for the first interval 67;50° and for the second interval 93;44°.

Then [see Fig. 10.7] let there be drawn in the plane of the ecliptic three equal circles: let the circle carrying the epicycle centre of Mars be ABG on centre D, the eccentre of uniform motion EZH on centre Θ, and the circle concentric with the ecliptic KLM on centre N, and let the diameter through all [three] centres be XOPR. Let A be the point at which the epicycle centre was at the first opposition, B the point where it was at the second opposition, and G the point where it was at the third opposition. Join ΘAE, ΘBZ, ΘHG, NKA, NLB and NGM. Then arc EZ of the eccentric [equant] is 81;44°, the amount of the first interval of mean motion, and arc ZH is 95;28°, the amount of the second interval. Furthermore arc KL of the ecliptic is 67;50°, the amount of the first interval of apparent motion, while arc LM is 93;44°, the amount of the second interval.

---

210   Ptolemy is referring to the crude periods of IX 3. Thus for Mars in 79 solar years occur 37 returns in anomaly and 42 returns in longitude. Assuming Ptolemy's year-length of 365;14,48d, one finds from this, for 4y 69d 20h, a longitudinal increment of 81;39°, and, for 4y 96d 1h, 95;23°. Using Ptolmy's procedure, and carrying out three iterations, I find from the above data 2e ≈ 11;57ᵖ, distance of 3rd opposition from perigee ≈ 44°. Comparison with Ptolemy's results from the more accurate data, 12ᵖ and 44;21°, shows that the differences are indeed negligible.

Now if arcs EZ and ZH of the eccentric [equant] were subtended by arcs KL and LM of the ecliptic, that would be all we would need in order to demonstrate the

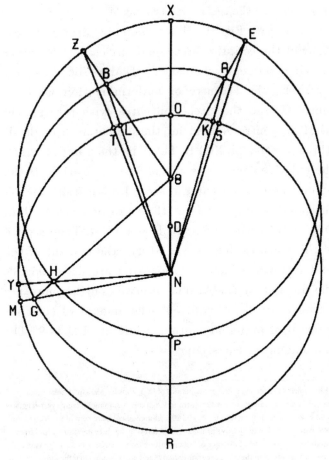

Fig. 10.7

eccentricity.[211] However, as it is, they[212] [arc KL and arc LM] subtend arcs AB and BG of the middle eccentre, which are not given; and if we join NSE, NTZ, NHY, we again find that arcs EZ and ZH of the eccentric [equant] are subtended by arcs ST and TY of the ecliptic, which are, obviously, not given either. Hence the difference arcs,[213] KS, LT and MY, must first be given, in order to carry out a rigorous demonstration of the ratio of the eccentricity starting from the corresponding arcs, EZ, ZH, and ST, TY. But the latter [arcs ST and TY] cannot be precisely determined until we have found the ratio of the eccentricity and [the position of] the apogee; however, even without the previous precise determination of eccentricity and apogee, the arcs are given approximately, since the difference arcs are not large. Therefore we shall first carry out the calculation as if the[214] arcs ST, TY did not differ significantly from the arcs KL, LM.

For [see Fig. 10.8] let the eccentre of mean motion of Mars be ABG, on which A is taken as the point of the first opposition, B of the second, and G of the third. Inside the eccentre take D as the centre of the ecliptic,

---

211 For the situation would be identical with that of the lunar hypothesis (IV 6).

212 Reading αὐται (with A,B [not reported by Heiberg]. Ar) for αὐταί at H324,8.

213 The arcs forming the differences between arc KL and arc TS, and between arc LM and arc TY.

214 Reading παρά τάς ΚΛΜ τῶν ΣΤΥ περιφερειῶν, at H324,22, for παρά τάς ΚΛΜ, περιφερειῶν ('as if arcs did not differ significantly from [arcs] KLM and STY', which is senseless). My text is the reading of all mss., Greek and Arabic. Heiberg omitted τῶν through a slip or a misprint. Because Manitius did not realize this, his translation here is badly flawed.

which is our point of view, draw in every case [where one has to carry out this kind of calculation] the lines joining the points of the three oppositions to the observer (as here AD, BD and GD), and, as a universal rule, produce one of the three lines so drawn to meet the circumference of the eccentre on the other side (as here GDE), and draw the line joining the other two opposition points (as in this case AB). Then, from the point where the straight line produced intersects the eccentre (as E), draw the lines joining it to the other two opposition points (as here EA and EB), and drop

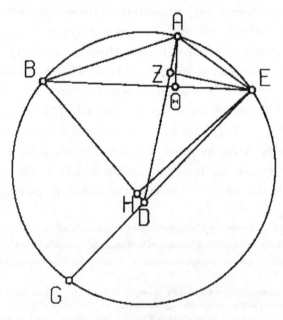

Fig. 10.8

perpendiculars [from the point corresponding to E] on to the lines joining the above-mentioned two points to the centre of the ecliptic (in this case, drop EZ on to AD, and EH on to BD). Also, drop a perpendicular from one of those two points on to the line joining the other with the extra point generated on the eccentre (as here, perpendicular AΘ on to line BE). If we always observe the above rules when drawing this type of figure, we will find that the same numerical ratios result however we decide to draw it.[215] The remainder of the demonstration will become clear as follows, on the basis of the above arcs for Mars.

Since arc BG of the eccentre is given as subtending 93;44° of the ecliptic, the angle at the centre of the ecliptic,

$$\angle \text{ BDG} = \begin{cases} 93;44° \text{ where 4 right angles} = 360° \\ 187;28°° \text{ where 2 right angles} = 360°° \end{cases}$$

and its supplement, ∠ EDH = 172;32°° in the same units.

Therefore, in the circle about right-angled triangle DEH,
arc EH = 172;32°

and EH = 119;45^P where hypotenuse DE = 120^p.

Similarly, since arc BG = 95;28°

the angle at the circumference, ∠ BEG = 95;28°° where 2 right angles = 360°°. But we found that ∠ BDE

---

215   I.e. whichever of the lines AD, BD, GD we decide to produce.

= 172;32°° in the same units.

Therefore the remaining angle [in triangle BDE],

∠ EBH = 92°° in the same units.

Therefore, in the circle about right-angled triangle BEH,

arc EH = 92°

and EH = 86;19^P where hypotenuse BE = 120^P.

Therefore where EH, as we showed, is 119;45^P, and ED = 120^P

BE = 166;29^P

Again, since the whole arc ABG of the eccentre is given as subtending [93;44° + 67;50° =] 161;34° of the ecliptic (the sum of both intervals),

∠ ADG = 161;34° where 4 right angles= 360°,

and, by subtraction [from 180°],

$$\angle \text{ ADE} = \begin{cases} 18;26° \text{ where 4 right angles} = 360° \\ 36;52°° \text{ where 2 right angles} = 360°°. \end{cases}$$

Therefore, in the circle about right-angled triangle

DEZ, arc EZ = 36;52°

and EZ = 37;57^P where hypotenuse DE = 120^P.

Similarly, since arc ABG of the eccentre is, by addition [of 81;44° to 95;28°], 177;12°.

∠ AEG = 177;12°° where 2 right angles = 360°°.

But we found that ∠ ADE = 36;52°° in the same units.

Therefore the remaining angle [in triangle ADE],

∠ DAE = 145;56°° in the same units.

Therefore, in the circle about right-angled triangle AEZ,

arc EZ = 145;56°

and EZ $= 114;44^P$ where hypotenuse AE $= 120^P$.
Therefore, where EZ, as was shown$= 37;57^P$ and ED$= 120^P$
$$AE = 39;42^P.$$

Again, since arc AB of the eccentre $= 81;44°$,

$\angle$ AEB $= 81;44°°$ where 2 right angles $= 360°°$.

Therefore, in the circle about right-angled triangle AEΘ,

arc AΘ $= 81;44°$

and arc EΘ $= 98;16°$ (supplement).

Therefore the corresponding chords

$$\left.\begin{array}{l} A\Theta = 78;31^P \\ \text{and } E\Theta = 90;45^P \end{array}\right\} \text{ where hypotenuse AE} = 120^P$$

Therefore where AE, as was shown, is $39;42^P$, and
DE is given as 120^P,

$$\Theta A = 25;58^P \text{ and } E\Theta = 30;2^P.$$

But the whole line EB was shown to be l $66;29^P$ in
the same units. Therefore, by subtraction, $\Theta B = 136;27^P$
where $\Theta A = 25;58^P$.

And $\Theta B^2 = 18615;16,$[216]

$$\Theta A^2 = 674;16,$$

so $AB^2 = \Theta B^2 + \Theta A^2 = 19289;32.$

∴ AB$= 138;53^P$ where ED $= 120^P$ and AE $= 39;42^P$.

But, where the diameter of the eccentre is 120^P. AB
$= 78;31^P$, since it subtends an arc of $81:44°$.

---

216 The square of 136;27 is 18618;36 to the nearest minute. The error has no significant effect on the size of AB below.

Therefore where AB = 78;31^P, and the diameter of the eccentre is 120^P,

ED = 67;50^P and AE = 22;44^P.

Therefore arc AE of the eccentre is 21;41°.[217]

And, by addition, arc EABG = [177;12° + 21;41° =] 198;53°. Therefore the remaining arc GE = 161;7°

and the corresponding chord GE = 118;22^P where the diameter of the eccentre is 120^P.

Now if GE had been found equal to the diameter of the eccentre, it is obvious that the centre would lie on GE, and the ratio of the eccentricity would immediately be apparent. But, since it is not equal [to the diameter], but makes segment EABG greater than a semi-circle, it is clear that the centre of the eccentre will fall within[218] the latter. Let it be at K [Fig. 10.9], and draw through D and K the diameter through both centres, LKDM, and drop perpendicular KNX from K on to GE.

Then, since, as we showed, EG = 118;22^P where diameter LM = 120^P,

and DE = 67;50^P in the same units,

by subtraction, GD = 50;32^P in the same units.

---

217 There are some serious errors here. For the chord AE one should find, from Ptolemy's figures, 22;27^P, and this is indeed the reading of Ger (but not the rest of the Arabic tradition) at H329,6. The arc of the latter, however, is not 21;41°, but 21;34°. Ptolemy's result (guaranteed by his further calculations), 21;41°, is the arc of 22;34^P. It looks as if the errors are Ptolemy's own (hence the reading of Ger is a misguided emendation). Did Ptolemy compute 22;27^P → 21;34°, and then, misreading his own notes, 22;34^P → 21;41°?

218 Reading ἐντός τούτου (with DG) at H329,17 for πρός τούτῳ ('at the latter'). Corrected by Manitius.

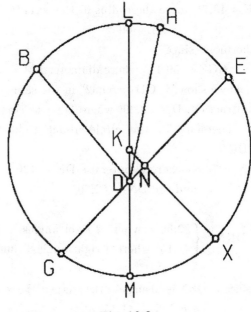

Fig. 10.9

Then, since ED.DG = LD.DM,[219]

LD.DM= [67;50 × 50;32 =] 3427;51.

But (LD.DM) + DK2 equals the square on half the whole line [LD + DM],[220] i.e. (LD.DM) + DK2 = LK2.

Now the square on the half is 3600, and (LD.DM) = 3427;51,

so DK^2 = 3600 − 3427;51 = 172;9,

and the distance between the centres,

---

219   Euclid III 35.

220   Euclid II 5.

DK $\approx$ 13;7^p where the radius of the eccentre,
KL= 60^{P221}.

Furthermore, since

GN = ¾GE = 59;11^P where diameter LM = 120^P,

and, as we showed, GD = 50;32^P in the same units,

by subtraction, DN= 8;39^p where DK was computed as 13;7^p. Therefore in the circle about right-angled triangle DKN,

DN = 79;8^p where hypotenuse DK = 120^p,

and arc DN = 82;30°.

$$\therefore \angle \text{DKN} = \begin{cases} 82;30°° & \text{where 2 right angles} = 360°° \\ 41;15° & \text{where 4 right angles} = 360°. \end{cases}$$

And since $\angle$ DKN is an angle at the centre of the eccentre,

arc MX = 41;15° also.

But the whole arc GMX = ½ arc GXE [= ½, 161;7°] = 80;34°.

Therefore, by subtraction, the arc from the third opposition to the perigee, arc GM = 39;19°.[222]

And it is obvious that, since arc BG is given as 95;28°,

by subtraction, the arc from the apogee to the second opposition, arc LB [= 180° – (95;28° + 39;19°)] = 45;13°,

and that, since arc AB is given as 81;44°,

by subtraction, the arc from the first opposition to the apogee, arc AL [= arc AB – arc LB] = 36;31°.

---

221 Accurate computation from Ptolemy's original data gives about 13;2½^P.

222 Accurate computation from Ptolemy's data gives 39;10°.

Taking the above quantities as given, let us investigate the differences which can be derived from them in the ecliptic arcs which we seek to determine at each of the oppositions [in turn]. Our investigation proceeds as follows.

[See Fig. 10.10.] From the previous figure [10.7] for the three oppositions let us draw separately the part representing the first opposition, draw the additional line AD, and drop perpendiculars DF and NQ from points D and N on to AΘ produced.

Then, since arc XE = 36;31°,

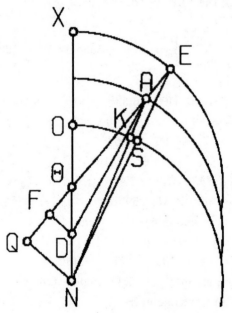

Fig. 10.10

$$\angle \; \text{E}\Theta\text{X} = \begin{cases} 36;31° \text{ where 4 right angles} = 360° \\ 73;2°° \text{ where 2 right angles} = 360°°. \end{cases}$$

And the vertically opposite angle $\text{D}\Theta\text{F} = 73;2°°$ in the same units also.

Therefore, in the circle about right-angled triangle $\text{D}\Theta\text{F}$,

$$\text{arc DF} = 73;2°$$
$$\text{and arc } \Theta\text{F} = 106;58° \text{ (supplement)}.$$

Therefore the corresponding chords

$$\left. \begin{array}{l} \text{DF} = 71;25^P \\ \text{and F}\Theta = 96;27^P \end{array} \right\} \text{where hypoteneuse D}\Theta = 120^P.$$

Therefore where $\text{D}\Theta = 6;33½^P$ and the radius of the eccentre, $\text{DA} = 60^P$,

$$\text{DF} = 3;54^P$$
$$\text{and F}\Theta = 5;16^P.$$

And since $\text{DA}^2 - \text{DF}^2 = \text{FA}^2$

$$\text{AF} = 59;52^P,$$

and, since $\text{QF} = \text{F}\Theta$,

by addition [of QF to FA], $\text{QA} = 65;8^P$

where $\text{NQ} = 2\text{DF} = 7;48^P$.

Hence hypotenuse [of right-angled triangle NAQ] $\text{NA} = 65;36^P$ in the same units.

Therefore, where $\text{NA} = 120^P$, $\text{NQ} = 14;16^P$, and, in the circle about right-angled triangle ANQ, arc NQ = 13;40°

∴ ∠ NAQ = 13;40°° where 2 right angles= 360°°.

Again, since QN was shown to be 7;48ᵖ and QΘ [= 2FΘ] to be 10;32ᵖ,

where the radius of the eccentre, ΘE = 60ᵖ,

by addition, QΘE = 70;32p in the same units,

and hence the hypotenuse [of right-angled triangle QNE] NE ≈ 71ᵖ in the same units.

Therefore, where NE = 120ᵖ, QN = 13;10ᵖ,[223] and, in the circle about right-angled triangle ENQ,

arc QN = 12;36°.

∴ ∠ NEQ = 12;36°° where 2 right angles= 360°°

But we found that ∠ NAQ = 13;40°°

in the same units.

Therefore, by subtraction [of ∠ NEQ from ∠ NAQ],

$$\angle \text{ANE} = \begin{cases} 1;4°° \text{ where 2 right angles} = 360°° \\ 0;32° \text{ where 4 right angles} = 360°. \end{cases}$$

That [0;32°], then, is the amount of arc KS of the ecliptic.

Next, draw a similar figure containing [the part of] the diagram for the second opposition [Fig. 10.11].

Then, since arc XZ is given as 45;13°,

---

223 The roundings here are particularly crude: from the immediately preceding numbers one finds NE = 70;57,48ᵖ, whence QN = 13;11,24ᵖ. Even NE = 71ᵖ leads to QN = 13;10,59ᵖ.

$$\angle \ X\Theta Z = \begin{cases} 45;13° \text{ where 4 right angles} = 360° \\ 90;26°° \text{ where 2 right angles} = 360°°, \end{cases}$$

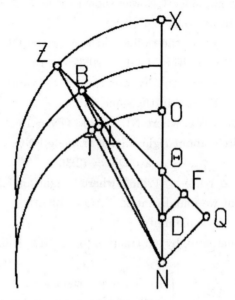

Fig. 10.11

and the vertically opposite angle DΘF $= 90;26°°$ in the same units, also.

Therefore, in the circle about right-angled triangle DΘF,

$$\text{arc DF} = 90;26°$$

and arc F$\Theta = 89;34°$ (supplement).

Therefore the corresponding chords

$$\left.\begin{array}{l} \text{DF} = 85;10^P \\ \text{and F}\Theta = 84;32^P \end{array}\right\} \text{where hypotenuse D}\Theta = 120^P.$$

228

Therefore where $D\Theta = 6;33\frac{1}{2}^p$ and the radius of the eccentre, $DB = 60^p$,

$$DF = 4;39^p \text{ and } F\Theta = 4;38^p.$$

And since $DB^2 - DF^2 = BF^2$

$$FB = 59;49^p,$$

and, since $FQ = F\Theta$,

by addition, $QB = 64;27$ p where NQ (= $2DF$) is computed as $9;18^p$. Therefore hypotenuse [of right-angled triangle NQB]

$NB = 65;6^{p224}$ in the same units. Therefore, where $NB = 120^p$, $NQ = 17;9^p$,

and, in the circle about right-angled triangle BNQ,

$$\text{arc } NQ = 16;26°$$

$\therefore \angle NBQ = 16;26°°$ where 2 right angles = $360°°$.

Again, since NQ was shown to be $9;18^p$, and $Q\Theta$ [= $2F\Theta$] = $9;16^p$, where the radius of the eccentre, $Z\Theta = 60^p$,

by addition; $Q\Theta Z = 69;16^p$ in the same units.

Hence hypotenuse NZ [of right-angled triangle NQZ] = $69;52^p$. Therefore, where hypotenuse $NZ = 120^p$, $NQ \approx 16^p$,

and, in the circle about right-angled triangle ZNQ,

$$\text{arc } NQ = 15;20°.$$

$\therefore \angle NZQ = 15;20°°$ where 2 right angles= $360°°$.

But we found that $\angle NBQ = 16;26°°$ in the same units.

---

224 Reading ξε̄; (with D,Ar) for ξθ̄ (69;6) at H335,9. The correction is assured by the preceding and subsequent computations.

Therefore, by
subtraction, $\angle$ BNZ $=\begin{cases} 1;6°° \text{ in the same units} \\ 0;33° \text{ where 4 right angles} = 360°. \end{cases}$

That [0;33°], then, is the amount of arc LT of the ecliptic.

Now, since we found arc KS as 0;32° for the first opposition, it is clear that the first interval, taken with respect to the eccentre,[225] will be greater than the interval of apparent motion by the sum of both arcs, [namely] 1;5°, and [hence] will contain 68;55°.

Then let [the part of] the diagram for the third opposition be drawn [Fig. 10.12]. Now, since arc PH is given as 39;19°,

$$\angle \text{ P}\Theta\text{H} = \begin{cases} 39;19° \text{ where 4 right angles} = 360° \\ 78;38°° \text{ where 2 right angles} = 360°°. \end{cases}$$

Therefore, in the circle about right-angled triangle DΘF, arc DF = 78;38°

and arc ΘF = 101;22° (supplement).

Therefore the corresponding chords

$$\left.\begin{array}{l} \text{DF} = 76;2^p \\ \text{and } \Theta\text{F} = 92;50^p \end{array}\right\} \text{ where hypotenuse D}\Theta = 120^p.$$

Therefore where the distance between the centres, DΘ = 6;33½^p, and the radius of the eccentre, DG = 60^p,

---

225   I.e. the equant: this is made explicit in XI I. See n.7.

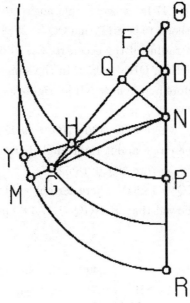

Fig. 10.12

DF= 4;9ᵖ and ΘF = 5;4ᵖ.

And since GD² – DF² = GF²

GF = 59;51ᵖ,

and, since ΘF = FQ

by subtraction, GQ = 54;47ᵖ where NQ(= 2DF) is computed as 8;18ᵖ.

Hence hypotenuse [of right-angled triangle NGQ]

NG = 55;25ᵖ in the same units. Therefore, where NG = 120ᵖ, NQ = 17;59ᵖ,

and, in the circle about right-angled triangle GNQ

arc NQ = 17;14°

$\therefore$ $\angle$ NGQ = 17;14°° where 2 right angles = 360°°. Again, since NQ was shown to be 8;18^p, and ΘQ [= 2FΘ] = 10;8^p,

where the radius of the eccentre, ΘH = 60^p,

by subtraction, QH = 49;52^p in the same units,

and therefore hypotenuse NH [of right-angled triangle NHQ] = 50;33^p.

Therefore, where NH= 120^p, NQ = 19;42^p,

and, in the circle about right-angled triangle HNQ

arc NQ = 18;54°.

$\therefore$ $\angle$ NHQ = 18;54°° where 2 right angles = 360°°.

But we showed that $\angle$ NGQ = 17;14°° in the same units.

Therefore by subtraction, $\angle$ GNH $= \begin{cases} 1;40°° \text{ in the same units} \\ 0;50° \text{ where 4 right angles} = 360°. \end{cases}$

That [0;50°], then, is the amount of arc MY of the ecliptic.

Now since we found arc LT as 0;33° for the second opposition, it is clear that the second interval, taken with respect to the eccentre, will be less than the interval of apparent motion by the sum of both arcs, [namely] 1;23°, and will [thus] contain 92;21°.

Using the ecliptic arcs thus computed for the two intervals, and, once more, the original arcs assumed for the eccentric [equant], and following the theorem demonstrated above for such elements, by means of which we determine [the position of] the apogee and the ratio of the eccentricity, we find (not to lengthen our account by going through the same [computations in detail again]), the distance between the centres, DK = 11;50ᴾ where the radius of the eccentre is 60ᴾ; the arc of the eccentre from the third opposition to the perigee, GM = 45;33°.[226]

Hence arc LB = [180° − (95;28° + 45;33°)] = 38;59°
and arc AL = [81;44° − 38;59°] = 42;45°.

Next, starting from these [arcs] as data, we found from our demonstration for each of the oppositions [separately] the following amounts for the true size of each of the arcs in question:

arc KS	0;28°
arc LT, about the same,	0;28°
and arc MY	0;40.[227]

We combined the [corrections] for the first and second oppositions, added the resulting 0;56° to the ecliptic arc of the first interval, 67;50°, and got the accurate interval with respect to the eccentre as 68;46°. Again, combining the [corrections] for the second and third oppositions, and subtracting the resulting 1;8° from the apparent motion on the ecliptic over the second interval, 93;44°, we got the accurate interval with respect to the eccentre as 92;36°.

---

226   From Ptolemy's elements, $\Delta\bar{\lambda}_1$= 81;44°, $\Delta\bar{\lambda}_2$ = 95;28°, $\Delta\lambda_1$ = 68;55°, $\Delta\lambda_2$ = 92;21°, I compute $2e$ = 11;50ᴾ, GM= 45;28°.

227   From a double eccentricity of 11;50ᴾ and Ptolemy's values for arcs GM, LB and AL, I find: arc KS = 0;27,49°, arc LT= 0;26,51°, arc MY= 0;39,31°.

Next, using the same procedure [as before], we determined a more accurate value for the ratio of the eccentricity and [the position of] the apogee; we found the distance between the centres, DK ≈ 12ᴾ where the radius of the eccentre, KL= 60ᴾ, arc GM of the eccentre = 44;21°,[228] whence, again, arc LB = 40; 11° and arc AL = 41;33°.

Next, we shall show by means of the same [configurations] that the observed apparent intervals between the three oppositions are found to be in agreement with the above quantities.

Let there be drawn [Fig. 10. 13] the diagram for the first opposition, but with only eccentre EZ, on which the epicycle centre is always carried, drawn in. Then

∠ AΘE = 41;33° where 4 right angles = 360°,

so where 2 right angles = 360°°,

∠ AΘE = 83;6°° = ∠ DΘF (vertically

opposite). Therefore, in the circle about right-angled triangle DΘF,

arc DF = 83;6°

and arc FΘ = 96;54° (supplement).

Therefore the corresponding chords

$$\left. \begin{array}{l} DF = 79;35^P \\ \text{and } FΘ = 89; 50^P \end{array} \right\} \text{ where hypotenuse } DΘ = 120^P.$$

---

228   From Ptolemy's elements I find: DK = 11;59,50ᴾ ≈ 12ᴾ, arc GM = 44;18,45° = 44;19°. Ptolemy is quite right to terminate his calculation here, since a further iteration produces a change in the eccentricity of less than 0;0,30ᴾ and in the line of the apsides of less than 5'.

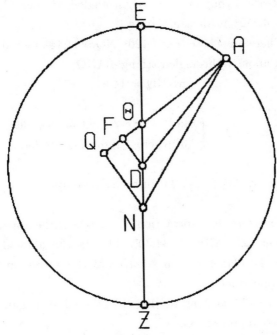

Fig. 10.13

Therefore where $D\Theta = 6^p$ and hypotenuse [of right-angled triangle DAF] $DA = 60^p$,

$$DF = 3;58^p$$

and $F\Theta = 4;30^p$

And since $DA^2 - DF^2 = FA^2$

$FA = 59;50^p$ in the same units.

Furthermore, since $F\Theta = FQ$ and $NQ = 2DF$,

by addition, $AQ = 64;20^p$ where $NQ = 7;57^p$.

Hence hypotenuse [of right-angled triangle NAQ] NA= 64;52ᵖ in the same units.

Therefore where NA = 120ᵖ, NQ = 14;44ᵖ, and, in the circle about right-angled triangle ANQ,

arc NQ = 14;6°.

$$\therefore \angle \text{NAQ} = \begin{cases} 14;6°° \text{ where 2 right angles} = 360°° \\ 7;3° \text{ where 4 right angles} = 360°. \end{cases}$$

But ∠ AΘE = 41;33° in the same units.

Therefore, by subtraction, the angle of the apparent position, ∠ ANE = 34;30°. This is the amount by which the planet was in advance of the apogee at the first opposition.

Let a similar diagram [Fig. 10.14] be drawn again for the second opposition.

Then the angle of the mean position of the epicycle,

∠ BΘE = 40;11° where 4 right angles = 360°, so where 2 right angles = 360°°,

∠ BΘE = 80;22°° = ∠ QΘN (vertically opposite).

Therefore, in the circle about right-angled triangle DΘF, arc DF = 80;22°

and arc FΘ = 99;38° (supplement).

Therefore the corresponding chords

$$\left. \begin{array}{l} \text{DF} = 77; 26ᵖ \\ \text{and FΘ} = 91; 41ᵖ \end{array} \right\} \text{where hypotenuse DΘ} = 120ᵖ·$$

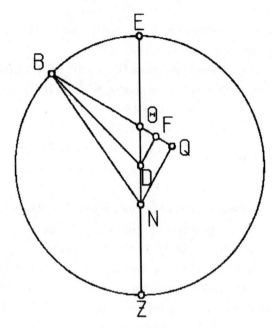

Fig. 10.14

Therefore where $D\Theta = 6^p$ and hypotenuse [of right-angled triangle DBF] $DB = 60^p$,

$$DF = 3;52^p$$

and $F\Theta = 4;35^p$. And since $DB^2 - DF^2 = BF^2$,

$$BF = 59;53^p \text{ in the same units.}$$

And, by the same argument [as before],[229]

since $F\Theta = FQ$, and $NQ = 2DF$,

by addition, $BQ = 64;28^p$ where $NQ = 7;44^p$.

---

229 Reading κατά ταύτά (as D, κατά τα αύτά, Ar) for κατά ταύτά ('according to this') at H342,23.

Hence hypotenuse [of right-angled triangle BNQ] BN = 64;56^p in the same units.

Therefore, where hypotenuse BN = 120^p, NQ= 14;19^p,[230] and, in the circle about right-angled triangle BNQ,

arc NQ= 13;42°.

$$\therefore \angle \text{ NBQ} = \begin{cases} 13;42°° \text{ where 2 right angles} = 360°° \\ 6;51° \text{ where 4 right angles} = 360°. \end{cases}$$

But $\angle$ BΘE = 40;11° in the same units. Therefore, by subtraction, the angle of apparent position,

$\angle$ ENB = 33;20° in the same units.

That [33;20°], then, is the amount by which the planet, in its apparent motion, was to the rear of the apogee at the second opposition. And we showed that at the first opposition it was 34;30° in advance of the apogee. Therefore the total distance [in apparent motion] from first to second opposition comes to 67;50°, in agreement with what we derived from the observations.

Let the diagram for the third opposition be drawn in the same way [Fig. 10.15]. In this case the angle of the mean position of the epicycle,

$$\angle \text{ GΘZ} = \begin{cases} 44;21° \text{ where 4 right angles} = 360° \\ 88;42°° \text{ where 2 right angles} = 360°°. \end{cases}$$

230 7;44 x 120/64;56 = I4;17,30, but if one carries out the above computations to 2 fractional sexagesimal places, one finds NQ = 14;18,41^p. As often, Ptolemy computed with greater accuracy than the text implies.

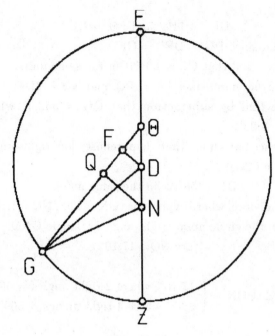

Fig. 10.15

Therefore, in the circle about right-angled triangle DΘF, arc DF = 88;42°

and arc FΘ = 91;18° (supplement).

Therefore the corresponding chords

$$\left.\begin{array}{l} DF = 83;53^P \\ \text{and } FΘ = 85;49^P \end{array}\right\} \text{where hypotenuse } DΘ = 120^P.$$

Therefore where DΘ = 6ᵖ and the radius of the eccentre, DG = 60ᵖ,

239

DF= 4;11½ᵖ and FΘ = 4;17ᵖ.

And since $DG^2 - DF^2 = GF^2$

we find that GF = 59;51ᵖ in the same units.

Furthermore, since FΘ = FQ, and NQ = 2DF,

we find by subtraction that QG = 55;34ᴾ where NQ = 8;23ᴾ.

Hence we find that hypotenuse [of right-angled triangle GNQ]

GN = 56;12ᵖ in the same units.

Therefore, where hypotenuse GN = 120ᵖ, NQ = 17;55ᵖ, and, in the circle about right-angled triangle GNQ,

arc NQ= 17;10°.

$$\therefore \angle \ \Theta GN = \begin{cases} 17;10°° \text{ where 2 right angles} = 360°° \\ 8;35° \text{ where 4 right angles} = 360°. \end{cases}$$

But ∠ GΘZ = 44;21° in the same units. Therefore, by addition, ∠ GNZ = 52;56° in the same units.

That [52;56°], then, is the amount by which the planet was in advance of the perigee at the third opposition. But we also showed that at the second opposition it was 33;20° to the rear of the apogee. So we have found 93;44° between the second and third oppositions, computed by subtraction [of the sum of 52;56° and 33;20° from 180°], in agreement with the amount observed for the second interval.

Furthermore, since the planet, when viewed at the third opposition along line GN, had a longitude of ♐

2;34° according to our observation, and angle GNZ at the centre of the ecliptic was shown to be 52;56°, it is clear that the perigee of the eccentre, at point Z, had a longitude of [♐ 2;34° + 52;56° =] ♑, 25;30°, while the apogee was diametrically opposite in ♋ 25;30°.

And if [see Fig. 10.16] we draw Mars' epicycle KLM on centre G and produce line ΘGM,[231] we will have, for the moment of the third opposition:

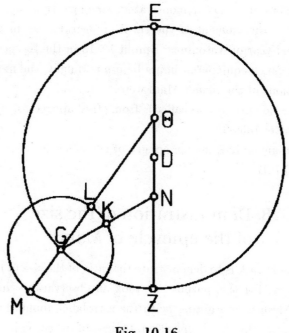

Fig. 10.16

---

231 Reading ΘΓΜ (with al-Ḥajjāj) for ΘΓ (ΘG) at H345,22.

mean motion of the epicycle counted from apogee of the eccentre: 135;39° (for its supplement, ∠ GΘZ, was shown to be 44;21°); mean motion of the planet from the epicycle apogee M (i.e. arc MK):171;25° (for ∠ ΘGN was shown to be 8;35° [above], and since it is an angle at the centre of the epicycle, the arc KL from the planet at K to the perigee at L is also 8;35°, hence the supplementary arc from the apogee M to the planet at K is, as already stated, 171;25°).

Thus we have demonstrated, among other things, that at the moment of the third opposition, i.e. in the second year of Antoninus, Epiphi 12/13 in the Egyptian calendar, 2 equinoctial hours before midnight, the mean positions of the planet Mars were:

in longitude (so-called) from the apogee of the eccentre: 135;39°

in anomaly from the apogee of the epicycle: 171;25°. Q.E.D.

8. Demonstration of the size of the epicycle of Mars[232]

Our next task is to demonstrate the ratio of the size of the epicycle. For this purpose we took an observation which we obtained by sighting [with the astrolabe] about three days after the third opposition, that is, in the second year

---

232   On the method employed here see *HAMA* 179–80, Pedersen 283–6.

of Antoninus, Epiphi [XI] 15/16 in the Egyptian calendar [139 May 30/31], 3 equinoctial hours before midnight. [That was the time,] for the twentieth degree of Libra [i.e ♎ 19°–20°] was culminating according to the astrolabe, while the mean sun was in ♊ 5;27° at that moment. Now when the star on the ear of wheat [Spica] was sighted in its proper position [on the instrument], Mars was seen to have a longitude of ♐ 1⅗°. At the same time it was observed to be the same distance (1; ⅗°) to the rear of the moon's centre. Now at that moment the moon's position was as follows:[233]

mean longitude	♐ 4;20°
true longitude	♍ 29;20°

(for its distance in anomaly from the epicycle apogee was 92°)

apparent longitude ♐ 0°,[234]

So from these considerations too the longitude of Mars was ♐1;36°, in agreement with the [astrolabe] sighting.

Hence, clearly, it was 53;54° in advance of the perigee.[235]

And the interval between the third opposition and this observation comprises

in longitude	about 1;32°
in anomaly	about 1;21°.[236]

---

233   These positions are computed (accurately), not for 9 p.m., but for 8:37 p.m., i.e. Ptolemy has applied the equation of time with respect to epoch as -23 minutes (it should be about -25 mins.)

234   Literally 'at the beginning of Sagittarius'.

235   Which was in ♑ 25;30°.

236   These mean motions do agree better with an interval of 2d 22h 3m than with one of 2d 23h (see n.57).

If we add the latter to the [mean] positions at the opposition in question[237] as demonstrated above, we get, for the moment of this observation:

distance of Mars in longitude from the apogee of the eccentre: 137;11°

distance in anomaly from the apogee of the epicycle: 172;46°.

With these elements as data, let [Fig. 10.17] the eccentric circle carrying the centre of the epicycle be ABG on centre D and diameter ADG, on which the centre of the ecliptic is taken at E, and the point of greater eccentricity [i.e. the equant] at Z. Draw the epicycle HΘK on centre B, draw ZKBH; EΘB and DB, and drop perpendiculars EL and DM from points D and E on to ZB. Let the planet be situated at point N on the epicycle, join EN, BN, and drop perpendicular BX from B on to EN produced.

Then, since the planet's distance from the apogee of the eccentre is 137;11°,

$$\angle\ BZG = [180° - 137;11° =]\begin{cases} 42;49° \text{ where 4 right angles} = 360° \\ 85;38°° \text{ where 2 right angles} = 360°°. \end{cases}$$

Therefore, in the circle about right-angled triangle DZM,
arc DM = 85;38°
and arc ZM = 94;22° (supplement).

237 Reading κατά τήν ύποκειμένην άκρώνυκτον (with D) for κατά τήν ύποκειμένην γ' άκρώνυκτον ('at the third opposition, which is the one in question') at H348,9–10.

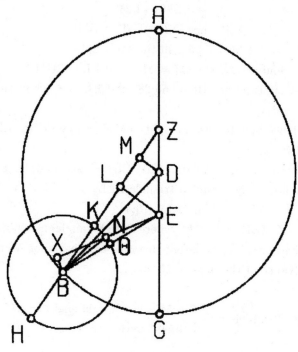

Fig. 10.17

Therefore the corresponding chords

$$\left.\begin{array}{l} DM = 81;34^P \\ \text{and } ZM = 88;1^P \end{array}\right\} \text{ where hypotenuse } DZ = 120^P.$$

Therefore where the distance between the centres, $DZ = 6^P$,
and the radius of the eccentre, $DB = 60^P$,

$$DM = 4;5^P$$

and ZM = 4;24^p.

And since DB^2 − DM^2 = BM^2

BM = 59;52^p in the same units.

Similarly, since ZM = ML, and EL = 2DM,

by subtraction, BL = 55;28^p and EL = 8;10^p in the same units.

Hence hypotenuse [of right-angled triangle EBL] EB = 56;4^p.

Therefore, where EB = 120^p, EL= 17;28^p, and, in the circle about right-angled triangle BEL,

arc EL= 16;44°

∴ ∠ ZBE = 16;44°° where 2 right angles = 360°°. Furthermore, the apparent distance of the planet Mars in advance of the perigee G,

∠ GEX is given as $\begin{cases} 53;54° \text{ where 4 right angles} = 360° \\ 107;48°° \text{ where 2 right angles} = 360°°. \end{cases}$

And, in the same units, ∠ ZBE = 16;44°° (shown above),

and ∠ GZB = 85;38°° (given),

so ∠ GEB = ∠ ZBE + ∠ GZB = 102;22°°.

Therefore, by subtraction [of ∠ GEB from ∠ GEX],

∠ BEX = 5;26°° in the same units,

and, in the circle about right-angled triangle BEX

arc BX = 5;26°.

So BX = 5;41^p where hypotenuse EB = 120^p. Therefore where EB, as was shown, = 56;4^p,

and the radius of the eccentre is 60ᵖ,
$$BX = 2;39ᵖ.$$

Similarly, since the distance of point N from the epicycle apogee H was 172;46°, and [hence], from the perigee K, 7;14°,

$$\angle \ KBN = \begin{cases} 7;14° \text{ where 4 right angles} = 360° \\ 14;28°° \text{ where 2 right angles} = 360°°. \end{cases}$$

But ∠ KBΘ was found as 16;44°° in the same units. Therefore, by subtraction, ∠ NBΘ = 2;16°°.

and, by addition, [of ∠ NBΘ to ∠ BEX], ∠ XNB = 7;42°°.

Therefore, in the circle about right-angled triangle BNX,
$$\text{arc } XB = 7;42°$$
and BX = 8;3ᵖ where hypotenuse BN = 120ᵖ.

Therefore where BX = 2;39ᵖ and the radius of the eccentre = 60ᵖ, the epicycle radius BN ≈ 39;30ᵖ.

Therefore the ratio of the radius of the eccentre to the radius of the epicycle is 60:39;30.

Q.E.D.

Bibliography

Apollonius, *Conics:* Apollonii Pergaei quae Graece exstant ed. I.L. Heiberg. 2 vols. Leipzig (Teubner), 1891, 1893.

Boll, Franz, 'Studien über Claudius Ptolemäus'. *Jahrbücher für Classische Philologie,* Supplementband 21, 1894, 51–244.

Calcidius: *Timaeus a Calcidio translatus commemarioque instructus.* ed. J.H. Waszink. London and Leiden, 1962.

Campanus: *Campanus of Novara and Medieval Planetary Theory: Theorica Planelarum,* ed. Francis S. Benjamin Jr. and G. J. Toomer. Madison, Wisconsin, 1971.

Diels-Kranz: *Die Fragmente der Vorsokratiker,* von Hermann Diels. Zehnte Auflage herausgegeben von Walther Kranz. 3 vols. Berlin, 1961.

Euclid: *Euclidis Elementa,* ed. I.L. Heiberg. (*Euclidis Opera Omnia,* vols. 1–5). Leipzig (Teubner), 1883–8.

Eudoxi Ars Astronomica qualis in charta Aegyptiaca superest, ed. Fr. Blass. Kiel [University Programme], 1887.

Geminus, *Eisagoge: Gemini Elementa Astronomiae*, ed. C. Manitius. Leipzig (Teubner), 1898.

HAMA: O. Neugebauer. *A History of Ancient Mathematical Astronomy*. 3 vols. Berlin-Heidelberg-New York, 1975.

Heiberg, J.L., see Apollonius, Euclid, Ptolemy, Theodosius.

Heron, *Dioptra:* Herons von Alexandria, Vermessungslehre and Dioptra, ed. Hermann Schöne. (*Heronis Alexandrini Opera quae supersunt omnia*, Vol. III). Leipzig (Teubner), 1903.

Hipparchus, *Comm, in Aral.: Hipparchi in Arati et Eudoxi Phaenomena Commentariorum*, Libri Tres rec. Carolus Manitius. Leipzig (Teubner), 1894.

Ideler, Ludwig, *Historische Untersuchungen über die astronomischen Beo-bachtungen der Alten.* Berlin, 1806.

Kugler, F.X., *Die Babylonische Mondrechnung.* Freiburg im Breisgau, 1900.

LSJ: *A Greek-English Lexicon* compiled by Henry George Liddell and Robert Scott. A New Edition Revised ... by Henry Stuart Jones. 2 vols. Oxford, 1940. Supplement, edited by E.A. Barber. Oxford, 1968.

Manitius [translation]: *Ptolemäus, Handbuch der Astronomie.* Deutsche Übersetzung von K. Manitius. 2 vols. Leipzig, 1912, 1913. Second edn., revised by O. Neugebauer. Leipzig, 1963.

Neugebauer [1]: O. Neugebauer, 'Untersuchungen zur antiken Astronomie V. Der Halleysche "Saros" und andere Ergänzungen zu UAA III'. *Quellen und Studien zur Geschichte der Mathematik, Astronomie und Physik* B 4, 1938, 407–11.

Neugebauer [2]; O. Neugebauer, 'On Some Aspects of Early Greek Astronomy'. *Proceedings of the American Philosophical Society.* Vol 116 no. 3, June 1972, 243–51.

Neugebauer, O. and Van Hoesen, H.B., *Greek Horoscopes.* (*Memoirs of the American Philosophical Society*, Vol. 48). Philadelphia, 1959.

Neugebauer, P.V., *Spezieller Kanon der Mondfinsternisse für Vorderasien und Ägypten von 3450 bis I v. Chr. Astronomische Abhandlungen.* (Ergänzungshefte zu den *Astronomischen Nachrichten*, Bd. 9 Nr. 2). Kiel, 1934.

Oppolzer, Th. v., *Canon der Finsternisse* (Ak. d. Wiss., Wien, Denkschriften, LII). Wien, 1887. English translation by Owen Gingerich as Canon of Eclipses. New York, 1962.

Pappus, *Commentary on the Almagest*; see Rome [1].

Pedersen: Olaf Pedersen, *A Survey of the Almagest.* (*Acta Historica Scientiarum Naturalium et Medicinalium*, Vol. 30). Odense University Press, 1974.

Peters, Christian H. F., and Knobel, Edward Ball, *Ptolemy's Catalogue of Stars. A Revision of the Almagest.* Washington, 1915.

Petersen, Viggo M., and Schmidt, Olaf, 'The Determination of the Longitude of the Apogee of the Orbit of the Sun according to Hipparchus and Ptolemy'. *Centaurus* 12, 1967, 73–96.

Price, Derek J., 'Precision Instruments to 1500', with a section on Hero's instruments by A.G. Drachmann. *A*

History of Technology, ed. Charles Singer et al., Vol. III. Oxford, 1957, 582–619.

Ptolemy, *Almagest: Claudii Ptolemaei Opera quae exstant omnia*. Vol. I, Syntaxis Mathematica, ed. J.L. Heiberg. 2 vols. Leipzig (Teubner), 1898, 1903.

Ptolemy, *Optics: L'Optique de Claude Ptolémée dans la version latine d'après l'arabe de l'émir Eugène de Sicile*, ed. Albert Lejeune. (Université de Louvain, Recueil de travaux d'histoire et de philologie, 4ᵉ série, fasc. 8). Louvain, 1956.

Ptolemy, *Planetary Hypotheses: The Arabic Version of Ptolemy's Planetary Hypotheses,* ed. Bernard R. Goldstein. *Transactions of the American Philosophical Society*, N.S. Vol. 57.4. Philadelphia, 1967.

Rome [1]: A. Rome (ed.), *Commentaires de Pappus et de Théon d'Alexandrie sur l'Almageste.*

Tome I. Pappus d'Alexandrie, *Commentaire sur les livres 5 et 6 de l'Almageste. (Studi e Testi 54).* Roma. 1931.

Tome II. Théon d'Alexandrie, *Commentaire sur les livres 1 et 2 de l'Almageste. (Studi e Testi 72).* Città del Vaticano, 1936.

Tome III. Théon d'Alexandrie, *Commentaire sur les livres 3 et 4 de l'Almageste. (Studi e Testi 106).* Città del Vaticano, 1943.

Rome [2]: A. Rome, 'Glanures dans l'idyiie 15 de Théocrite'. Academie royale de Belgique, *Bulletin de la Classe des Lettres*, 5ᵉ ser. T. 37, 1951, 260–7.

Rome [3]: A. Rome, 'Les Observations d'Equinoxes de Ptolémée. Ptolémée et le mouvement de l'apogée solaire'. *Ciel et Terre* 59, 1943, 1–15.

Rome [4]: A. Rome, 'L'Astrolabe et le Météoroscope d'après le commentaire de Pappus sur le 5e livre de l'Almageste'. *Annales de la Société Scientifique de Bruxelles* 47, 1927, Deuxième partie, Mémoires, 77–102.

Sawyer, Frederick W., 'On Ptolemy's Determination of the Apsidal Line for Venus'. Appendix to: Bernard R. Goldstein, 'Remarks on Ptolemy's Equant Model in Islamic Astronomy'. *ΠΡΙΣΜΑΤΑ* (Festschrift für Willy Hartner), ed. Y. Maeyama and W.G. Saltzer. Wiesbaden, 1977, 169–81.

Schmidt, Olaf, 'Bestemmelsen af Epoken for Maanens Middelbevaegelse i. Bredde hos Hipparch og Ptolemaeus'. *Matematisk Tidsskrift B* 1937, 27–32.

Swerdlow, N., and Neugebauer, O., *Mathematical Astronomy in Copernicus' De Revolutionibus. (Studies in the History of Mathematics and Physical Sciences)*. 1984. Springer New York, NY, 2012.

Theodosius, *Sphaerica:* Theodosius Tripolites, *Sphaerica*, von J.L. Heiberg. (Abh. der Ges. d. Wiss. zu Göttingen, Phil-hist. Kl., N.F. XIX.3). Berlin, 1927.

Théon of Alexandria, *Commentary on the Almagest*: see Rome [1]

Théon of Smyrna: *Théonis Smyrnaei Expositio rerum mathematicarum ad legendum Platonem utilium*, ed. E. Hiller. Leipzig (Teubner), 1878.

Toomer [1]: G.J. Toomer, 'The Mathematician Zenodorus'. *Greek, Roman and Byzantine Studies* 13, 1972, 177–92.

Toomer [2]: G.J. Toomer, 'The Chord Table of Hipparchus and the Early History of Greek Trigonometry'. *Centaurus* 18, 1973, 6–28.

Toomer [3]: G.J. Toomer, review of Olaf Pedersen, *A Survey of the Almagest*, Archives Internationales d'Histoire des Sciences 27, 1977, 137–50.

Toomer [5]: article PTOLEMY. *Dictionary of Scientific Biography* XI, 1975, 186–206.

Toomer [8]: G.J. Toomer, 'The Size of the Lunar Epicycle According to Hipparchus'. *Centaurus* 12, 1967, 145–50.

Toomer [9]: G.J. Toomer, 'Hipparchus on the Distances of the Sun and Moon'. *Archive for History of Exact Sciences* 14, 1974, 126-42.

Toomer [11]: G.J. Toomer, 'Hipparchus' Empirical Basis for His Lunar Mean Motions'. *Centaurus* 24, 1981, 97–109.

Biographies

Claudius Ptolemy

The Almagest

Claudius Ptolemy (about 100–170 CE) lived in Alexandria, Egypt, part of the Roman Empire. A mathematician, geographer and astrologer, his famous book on astronomy codified the ancient view of the universe, with the sun and observable planets orbiting around the earth.

Christián C. Carman

The Introduction to The Book of Astronomy in Antiquity

Christián C. Carman is professor and researcher at the Universidad Nacional de Quilmes, Argentina, and research member of the National Research Council of Argentina (CONICET). He works on topics related to philosophy of science as well as history of ancient astronomy (mainly related to the Antikythera Mechanism, Aristarchus of Samos and Ptolemy), and history of Early Modern astronomy.

Marika Taylor

Series Foreword

Professor Marika Taylor is a Professor of Theoretical Physics and Head of School within Mathematical Sciences at the University of Southampton. Her research interests include all aspects of string theory, gravitational physics and quantum field theory. In recent years much of her work has been focused on holographic dualities and their implications. Marika's research has featured in such publications as *Physical Review, Journal of High Energy Physics* and *General Relativity and Gravitation* among others.

With introductions and a new foreword.

A new series of accessible books bringing
together ancient, medieval and modern texts
in a concise form. Each title features an
introduction placing the book in context and
highlighting its special contribution to the
advancement of human understanding.

See our expanding range at
flametreepublishing.com